Real Analysis

Gary Towsley

Version Date: March 31, 2021

ISBN: 978-1-942341-74-1

Published by Milne Library
1 College Circle
Geneseo, NY 14454
publishing@milne-library.org

Acknowledgements

First, I would like to thank the many students I have taught in Real Analysis at the State University of New York at Geneseo for their patience while this text evolved from a few pages of statements of definitions and theorems to a complete text. I would also like to thank the Real Analysis students at the State University of New York at Oswego who used the book in the course taught by Professor Elizabeth Wilcox. She brought back many helpful comments from her students.

Then I would like to offer a very large thank you to Professor Wilcox for all the work she has put in editing and improving the text. There is not one page that her reading and insights has not improved.

Contents

Preface

This book is simply a text derived from the lectures I have given in our Real Analysis course over the last forty-four years. The course follows the recommendations in the 2015 Curriculum Guide from the Mathematical Association of America. This text, itself, is a sketch of the course, including all the theorems and definitions but often giving a less than full exposition of the topics.

Introduction: A Quick History of the Development of Real Analysis

The main subject matter of Real Analysis is the set of real numbers and the properties of functions from the real numbers to the real numbers. This subject has been developing for many thousands of years in many different societies from many different regions of the Earth. All ancient civilizations had methods for measuring lengths, areas, and volumes. They applied discrete numbers to continuous quantities. "A log was 6 cubits long." We are fortunate to have had the discussion of this mix of algebra and geometry in Ancient Greece preserved for us. For example, Aristotle spends considerable time discussing many aspects of lines. He discussed lines in terms of two sets of opposites — discrete vs. continuous and abstract vs. physical. The counting numbers $\{1, 2, 3, \dots\}$ form a discrete set while a line segment is continuous. One way of thinking about the distinction between the discrete and the continuous is by thinking about counting and measuring. One counts the number of olives in a crate of olives while one measures the volume of olive oil extracted from the crate. He also distinguished between physical lines — for example a curve drawn on a piece of wood — and the kind of abstract line that exists only in our minds. Trying to determine the relation between the discrete and the continuous he asked the questions — "Is a line composed of points? Is there something else needed beyond points to make it a line?"

Aristotle worked on the relations of points on a line and the whole line. He knew that a point on a line divided the line into two parts. He also knew that if two points were specified on a line then there was always a point on the line between those two points. But could one divide a line up into points so that there was nothing left beyond a collection of points? His answer was interesting. He said that a line was potentially infinitely divisible by points. That is, between any two points on the line there was always a third point. But a line was not actually infinitely divisible by points. Thus his answer to the basic question was that the line could not in actuality be reduced to a set of points.

Greek mathematicians worked on abstract lines and looked for locations on those lines. The locations were points. Given a line one could pick two distinct points on the line and call the segment between them a unit, that is a segment of length one. One could then try to measure the length of an arbitrary segment in terms of this unit length. It is easy to see that this wouldn't work in general. Take the unit segment and cut it at some interior point. Clearly the two resulting segments were less than one in length. To measure lines numerically the ancients made an assumption. Given a unit length and any other segment there were two whole numbers m and n such that m copies of the unit segment was exactly the same length as n copies of the other segment. Call the unit segment I and the other segment A. They expressed the length of A with the proportion $I : A = n : m$. We would say that the length of A was m/n units.

$$I : A = n : m$$
$$\frac{I}{A} = \frac{n}{m}$$
$$A = \left(\frac{m}{n}\right) I$$

The Greeks did not use either of the fractional forms above, just the proportion. Their assumption that such whole numbers m and n always existed led to the notion of commensurability. They assumed that any two line segments were "commensurable," that is they had a common measure, a smaller segment that "measured" each of the original. Suppose that $I : A = 2 : 3$. In terms of the unit length (the length of I), A has length $3/2$. A segment M of length exactly half of I would measure I. That is $I = 2M$, while exactly three copies of M would equal A in length. Thus M would be a common measure of the segments I and A. It fell to the Pythagoreans somewhere around the year 425 B.C.E. to discover that the assumption of commensurability was false. If S and D are respectively the side and diagonal of a square then there did not exist whole numbers m and n such that $S : D = m : n$. We express this by the statement that the square root of two is irrational, in other words the ratio of side to diagonal is not a ratio of whole numbers.

This discovery led to deep research into Geometry and its relation to Arithmetic culminating in Book X of Euclid's Elements. From Aristotle's point of view the discovery meant that there were many more points on a line than previously thought. Let our unit segment I be represented by AB where A and B are the endpoints of I. Then there is a point C somewhere on the line such that $AB : AC = S : D$, where S and D are the side and diagonal of a square. How many more points were there on a line? That question waited for about 2200 years for an answer. In the late 1800s Georg Cantor developed Set Theory and the Theory of Infinite Cardinal Numbers. He showed that the number of points on a line segment was a bigger infinity than the infinity of rational numbers. Cantor's discovery led to the realization that lines were more mysterious than Aristotle had imagined. It is this mystery that is the subject matter of Real Analysis. At approximately the same time that Cantor was working other mathematicians were looking at both the algebraic properties and the space-like properties of lines. This all came together in the description of the real numbers as a complete, ordered field. That is where we will begin the course.

Chapter 1

The Natural Numbers, the Rational Numbers and their Arithmetic

One of the goals of this book is to explore the real numbers. The set, $\mathbb{R}$, can be defined from the rational numbers though we will only do this as a sketch in the appendix. The rational numbers, $\mathbb{Q}$, are defined via the integers. The set of integers, $\mathbb{Z}$, is based on the set of natural numbers. We will start by considering the natural numbers, which we denote by the symbol $\mathbb{N}$. This is the set $\mathbb{N} = \{1, 2, 3, 4, \dots\}$. Some mathematicians include 0 in $\mathbb{N}$. It is a matter of choice with no real importance whether 0 is or is not in $\mathbb{N}$. We just assume that it is not. We have claimed that the real numbers can be defined but this is a long process. We will indicate in detail only the first step, namely the axiomatic definition of the natural numbers. After that we will simply assume the relevant arithmetic facts.

The natural numbers are the first numbers you learned about and they are basic to everything. It may seem surprising but the natural numbers were axiomatized in the year 1889 by Giuseppe Peano. That is, Peano stated a set of axioms that in a certain sense defined the natural numbers. We have reduced his set of axioms to five by combining some of the originals.

Axiom 1. 1 is a natural number.

Each natural number n has a successor, namely $n + 1$. We denote its successor by $n + +$ for the following axiom.

Axiom 2. If n is a natural number then $n + +$ is a natural number.

The idea of successor is the basic idea involved in counting. It is really the primitive intuition that we develop in kindergarten that makes numbers make sense.

1 is a special natural number as shown in the next axiom.

Axiom 3. 1 is not the successor of any natural number.

Axiom 4. If m and n are distinct natural numbers ($m \neq n$), then $m + + \neq n + +$.

What allows us to put all of the arithmetic of the natural numbers on a sound footing is the principle of induction. We take that principle as an axiom.

Axiom 5. Let $P(n)$ be a statement about the natural number n. If $P(1)$ is true and if whenever $P(n)$ is true then $P(n + +)$ is also true, then $P(n)$ is true for all natural numbers n.

Definition 1.1 (The Set of Natural Numbers)**.** The set $\mathbb{N}$ of natural numbers consists of 1 and all the successors of any element in $\mathbb{N}$.

An equivalent form of Axiom 5 sometimes makes it easier to understand induction.

Definition 1.2 (Well-Ordered Set). A set A is *well-ordered* if every non-empty subset of A has a smallest element.

Axiom 5 (Alternate). The set of natural numbers is well ordered.

We will prove that the alternate axiom implies Axiom 5, which should make clearer the connection between induction and the intuitive idea of the natural numbers.

Theorem 1.3. Axiom 5 (Alternate) implies Axiom 5.

To prove Theorem 1.3, let P be a statement about the natural number n. We'll show that if P satisfies the following two properties, then P is true for all natural numbers:

i $P(1)$ is true,

ii if $P(k)$ is true, then $P(k+1)$ is also true.

Proof. Let $P(n)$ be a statement about the natural numbers that satisfies i) and ii) but is not true for every natural number n. Let $S = \{n \in \mathbb{N} \mid P(n) \text{ is false}\}$. Then by our assumption S is not empty. By the Alternate Axiom 5, S has a smallest element. Call it m. The statement $P(m)$ is false since $m \in S$. Thus by i) $m \neq 1$. Since m is a natural number this implies that $m > 1$. Thus $k = m1$ is a natural number smaller than m. Hence $P(k)$ is true since $m = k + 1$ is the smallest natural number for which P is false. But by ii) $P(k+1) = P(m)$ is true which contradicts the assumption that S is non-empty. □

Summing this up we see that induction works because if a statement has a counterexample it has to have a smallest counterexample. Axiom 5 says that the conditions on P imply that it has no smallest counterexample.

We now expand our work to the set of integers denoted by $\mathbb{Z} = \{\ldots, -3, -2, -1, 0, 1, 2, 3, \ldots\}$. While mathematical induction will be needed directly in the text only rarely, it has very important uses. There are many variants to the definition of induction given above. Essentially induction is used to prove that a statement about integers is true for all integers in certain kinds of sets. The form given in the outline assumes that $P(n)$ is a statement about the natural number n. The base case is $n = 1$, that is $P(1)$ is the first statement that must be shown true. However the principle of induction can begin at any integer z_0. In this form the principle reads:

> If $P(n)$ is a statement about the integer n and the following two conditions hold, then $P(n)$ is true for all integers in the set $\{n \in \mathbb{Z} \mid z_0 \leq n\}$. The conditions are:
>
> i $P(z_0)$ is true.
>
> ii For any integer k satisfying $k \geq z_0$, if $P(k)$ is true then $P(k+1)$ is true.

A useful exercise is to state the principle of induction for sets of the form $\{n \in \mathbb{Z} \mid n \leq z_0\}$.

Now we consider the rational numbers. They will serve a very important role in this course. They will be the skeleton over which we build the real numbers. The set of rational numbers, or $\mathbb{Q}$, is defined by $\mathbb{Q} = \{\frac{p}{q} \mid p, q \in \mathbb{Z}, q > 0, \gcd(p,q) = 1\}$, where $\gcd(p,q)$ represents the greatest common divisor of p and q. In this definition we pick a particular form for each rational number, namely as a quotient of integers in lowest terms with a positive denominator. This definition helps us avoid certain computational difficulties.

Example 1.4. We know that $\sqrt{\frac{a}{b}} = \frac{\sqrt{a}}{\sqrt{b}}$ and that $-1 = \frac{-1}{1} = \frac{1}{-1}$. Combining these yields

$$\sqrt{\frac{-1}{1}} = \frac{\sqrt{-1}}{1} = \sqrt{-1} = \sqrt{\frac{1}{-1}} = \frac{1}{\sqrt{-1}}.$$

However, the complex numbers $\sqrt{-1}$ and $\frac{1}{\sqrt{-1}}$ are not equal to each other, but are negatives of each other. Our definition of the rationals avoids this problem.

We make a connection between the rational numbers and the real numbers through decimal expansions. The process of long division (with potentially infinitely many steps) gives a decimal expansion for each rational number.

Example 1.5.

$$\begin{aligned}
\frac{1}{2} &= 0.5 = 0.5000\cdots = 0.4999\cdots \\
\frac{1}{3} &= 0.333\cdots \\
\frac{1}{7} &= 0.142857142857\cdots \\
\frac{7}{40} &= 0.175 = 0.1750000\cdots = 0.174999\cdots
\end{aligned}$$

We have used $\cdots$ to indicate the repeating pattern of digits in each case. There is a better way to do this using a bar notation.

Example 1.6.

$$\begin{aligned}
\frac{1}{2} &= 0.5 = 0.4\overline{9} \\
\frac{1}{3} &= 0.\overline{3} \\
\frac{1}{7} &= 0.\overline{142857}\cdots \\
\frac{7}{40} &= 0.175 = 0.174\overline{9}
\end{aligned}$$

Theorem 1.7. Every rational number has a repeating decimal expansion.

Example 1.8. The rational number $\frac{1}{2}$ has two distinct repeating decimal expansions, namely $\frac{1}{2} = 0.5 = 0.5\overline{0} = 0.4\overline{9}$. The first expansion is called a terminating expansion since it is all zeros after some point in the expansion. The second expansion is said to be repeating and non-terminating. Are these two expansions really the same number? They differ in infinitely many places. That $\frac{1}{2} = 0.5$ is easy to show by long division but what about that $\frac{1}{2} = 0.4\overline{9} = 0.49999\cdots$?

Let $x = 0.49999\cdots$. Clearly (or not so clearly) $10x = 4.999999\cdots$. Subtracting yields

$$10x - x = 9x = 4.99999\cdots - 0.4999999\cdots = 4.5000\cdots.$$

Thus $9x = 4.5$ or $x = \frac{4.5}{9} = \frac{1}{2}$.

Example 1.9. The number $x = 17.3571428571428\cdots = 17.3\overline{571428}$ is a repeating, non-terminating decimal expansion. It also represents a rational number. Notice that the expansion repeats in 6 places. Multiply x by one million or 10^6.

$$\begin{aligned} 1,000,000x &= 17,357,142.8\overline{571428} \\ x &= 17.3\overline{571428} \end{aligned}$$

Subtracting yields $999,999x = 17,357,125.5$. Note that the repeating part matches up in position so that most of the subtractions result in 0. Now we have $x = \frac{17,357,125.5}{999,999} = \frac{34,714,251}{1,999,998} = \frac{243}{14}$. The last fraction is in lowest terms.

Which rational numbers have terminating decimal expansions? The denominator when the fraction is expressed in lowest terms tells the full story. A terminating decimal expansion can always be written as a fraction with a denominator which is a power of 10 (and an integer numerator).

Example 1.10.

$$12.345 = \frac{12,345}{1,000} = \frac{2,469}{200}$$

Thus in lowest terms a fraction which has a terminating decimal expansion has a denominator which is a factor of a power of 10. In other words, fractions whose denominators are products of only 2's and 5's have terminating expansions.

We have made all the above computations in base 10 arithmetic but the same facts are true in any other base 2 or greater.

For example as a base 2 or binary expansion $\frac{1}{2} = 0.1_2$ and $\frac{3}{16} = 0.0011_2$. Both of these fractions have denominators which are powers of 2 and are hence terminating. The situation is different for a fraction like $\frac{1}{3}$.

$$\frac{1}{3} = 0.010101\cdots_2 = 0.\overline{01}_2$$

We will wait until we cover infinite series to see how to generate this expansion.

Example 1.11. A few examples of expansions in other bases.

$$\begin{aligned} \frac{1}{2} &= 0.1111\cdots_3 \\ \frac{1}{2} &= 0.2222\cdots_5 \\ \frac{1}{2} &= 0.3333\cdots_7 \\ \frac{1}{3} &= 0.131313\cdots_5 \\ \frac{1}{3} &= 0.2222\cdots_7 \end{aligned}$$

Example 1.12 (Dirichlet's Function). This is a very simply defined function that will turn out to be useful in several different contexts. Let $d : \mathbb{R} \to \mathbb{R}$ be defined by:

$$d(x) = \begin{cases} 0 & \text{if } x \notin \mathbb{Q} \\ 1 & \text{if } x \in \mathbb{Q} \end{cases}.$$

Here are a few values of the function: $d(0) = 1$, $d(7) = 1$, $d\left(\frac{1}{\sqrt{2}}\right) = 0$, $d(3 + \sqrt[3]{7}) = 0$.

The graph of d would look roughly like two horizontal lines: $y = 0$ and $y = 1$. They would not be solid lines, however.

Example 1.13 (Thomae's Function)**.** Thomae's function is related to Dirichlet's function, defined above but it is a bit more complicated to compute. We assume in our definition that every rational number is expressed as $\frac{p}{q}$, where p and q share no common factors greater than 1 and that $q > 0$. Given this we define Thomae's function, $t : \mathbb{R} \to \mathbb{R}$ by

$$t(x) = \begin{cases} 0 & \text{if } x \notin \mathbb{Q} \\ \frac{1}{q} & \text{if } x = \frac{p}{q} \in \mathbb{Q} \end{cases}.$$

Let n be any integer. Then expressed as a rational number of the correct form $n = \frac{n}{1}$. Hence $t(n) = 1$. Thus $t(0) = 1$, $t(-7) = 1$, and $t(1000) = 1$. Additionally,

$$t(\sqrt{2}) = t(\sqrt[3]{7}) = t(1 - \sqrt{3}) = 0,$$

while $t\left(\frac{-3}{8}\right) = \frac{1}{8}$ and

$$t(1.024) = t\left(\frac{1024}{1000}\right) = t\left(\frac{128}{125}\right) = \frac{1}{125}.$$

1.1 Exercises

Exercise 1.1. Express the following rational numbers as repeating decimal expansions: $\frac{3}{11}$, $\frac{5}{9}$, $\frac{20}{13}$, and $\frac{6}{37}$.

Exercise 1.2. Express the following repeating decimal expansions as quotients of integers in lowest terms: $0.123123\cdots$, $0.1231212\cdots$, and $0.699999\cdots$.

Exercise 1.3. An irrational number can be defined as a number represented by an infinite decimal expansion that is neither repeating nor terminating. It is rather difficult to prove that commonly used irrational numbers like $\sqrt{2}$, $\sqrt[3]{7}$, π, or e have decimal expansions that are neither repeating nor terminating. But one can define an irrational number by creating a non-repeating, non-terminating decimal expansion. Create a non-terminating, non-repeating decimal expansion, hence create an irrational number.

Exercise 1.4. Let $x = 0.137137137\cdots$ and $y = 0.93939393\cdots$. Express $x + y$ as a repeating decimal expansion.

Exercise 1.5. The decimal expansions $0.\overline{142857}$ and $0.\overline{10102}$ clearly represent rational numbers. Thus their sum is a rational number. In how many places does its decimal expansion repeat?

Exercise 1.6. Prove that if $x > 0$ and n is a natural number then $(1 + x)^n \geq 1 + nx$.

Exercise 1.7. Prove that the sum of the first n consecutive odd numbers starting at 1 is n^2.

Exercise 1.8. Prove that if p is a prime number then for every $n \in \mathbb{N}$, the number $n^p - n$is a multiple of p.

Exercise 1.9. Let the sequence $\{x_n\}$ be defined as follows: $x_1 = 1$, $x_{n+1} = \frac{x_n}{2} + 1$ for all $n \in \mathbb{N}$. Using induction, prove that the sequence is increasing and bounded above by 2.

Exercise 1.10. Let A be a set with exactly n elements. Prove that A has exactly 2^n different subsets.

Exercise 1.11. For each $n \in \mathbb{N}$, let $I_n = \{1, 2, 3, \ldots, n\}$. Prove by induction that if $n \neq m$ then I_n and I_m cannot be put into 1-1 correspondence.

Exercise 1.12. Let $\{F_n\}$ be the sequence of Fibonacci numbers, that is:

$$F_0 = 0,\ F_1 = 1,\ \text{ and } F_{n+2} = F_n + F_{n+1} \text{ for all } n \geq 0.$$

Thus $\{F_n\} = \{0, 1, 1, 2, 3, 5, 8, 13, 21, \ldots\}$.

Prove the following identities.

a) $F_0 + F_1 + \cdots + F_n = F_{n+2} - 1$

b) $F_0 + F_2 + F_4 + \cdots + F_{2n} = F_{2n+1} - 1$

c) $F_1 + F_3 + F_5 + \cdots + F_{2n-1} = F_{2n}$

Exercise 1.13. Find a formula for the sum of the first n perfect cubes, starting from 1. Use induction to prove that your formula is correct.

Chapter 2

Preliminaries Concerning Sets and Functions

Real analysis is often described as the course in which one proves the results one uses in Calculus. This is true as far as it goes but more needs to be said. First, real analysis describes as fully as possible the set called the real numbers, $\mathbb{R}$. The real numbers have both an algebraic structure and a space structure. The algebraic structure consists of the operations on the set and the variety of kinds of numbers (rational and irrational, etc.) contained in the set. The space structure shows how points are related to nearby points. Much of what is developed in this course comes about through the intertwining of the two structures.

There are many ways to describe the real numbers. One is simply that the real numbers are all the numbers that can be expressed as decimal expansions. This definition needs a caution in that many real numbers have two different decimal expansions. For example 1 and $0.999999\cdots$ are the same number. Any rational number whose denominator is a product of 2's and 5's only similarly has two decimal expansions. For example $\frac{3}{40} = \frac{75}{1{,}000} = 0.075 = 0.749999\cdots$.

Another way to describe the real numbers is to start with the rational numbers, $\mathbb{Q}$, and to fill in the holes between these numbers using an axiom called the Completeness Axiom. The decimal expansions of a real number provide a sequence of rational numbers that tend to that real number. Think about how you would start with a decimal expansion like $1.123123123\cdots$ and produce a sequence of rational numbers that converges to that number.

Perhaps the best way to talk about the real numbers initially is to describe them as a complete ordered field. We define first a field and then an ordered field. The completeness aspect will appear in chapter 3.

A field is essentially a set of numbers on which one can do ordinary algebra.

Definition 2.1 (Field)**.** Let F be a set with two binary operations, $+$ and $\times$ (or $\cdot$), that satisfy the following axioms. Then F is called a *field*. For each of the following assume that $x, y, z \in F$.

a) (Closure of Addition) For all x and y in F, $x + y \in F$.

b) (Commutativity of Addition) For all x and y in F, $x + y = y + x$.

c) (Associativity of Addition) For all x, y, and z in F, $(x + y) + z = x + (y + z)$.

d) (Identity Element for Addition) There is an element 0 in F such that $0 + x = x + 0 = x$ for every $x \in F$.

e) (Additive Inverse of an Element) To each $x \in F$ there corresponds a $-x \in F$ such that $x + (-x) = (-x) + x = 0$.

f) (Closure of Multiplication) For all x and y in F, $x \times y \in F$.

g) (Commutativity of Multiplication) For all x and y in F, $x \times y = y \times x$.

h) (Associativity of Multiplication) For all x, y, and z in F, $(x \times y) \times z = x \times (y \times z)$.

i) (Identity Element for Multiplication) There is an element 1 in F with $1 \neq 0$ such that $1 \times x = x \times 1 = x$ for every $x \in F$.

j) (Multiplicative Inverses) If $x \neq 0$ is in F, then there exists $\frac{1}{x} \in F$ satisfying $x \times \frac{1}{x} = \frac{1}{x} \times x = 1$.

k) (Distributivity of Multiplication over Addition) For all x, y, and z in F, $(x + y) \times z = (x \times z) + (y \times z)$.

These axioms guarantee that we can perform all the operations we learned in algebra with the elements of a field. Note that several necessary algebraic facts are not part of the axioms and need to be derived. We have not assumed, for instance that additive inverses and multiplicative inverses are unique. We have not assumed that 0 and 1 are distinct. There are many other results that one uses without thinking in algebra, including those listed below.

1. $0 \cdot x = 0$
2. $(-1) \cdot x = -x$
3. $(-x) \cdot y = -(x \cdot y)$
4. $(-x) \cdot (-y) = x \cdot y$

We will not prove these results but we will use them quite freely.

Definition 2.2 (Ordered Field)**.** An ordered field is a field F with a binary relation $<$ called an order satisfying:

a) If $x, y, z \in F$ and $x < y$ then $x + z < y + z$.

b) If $0 < x$ and $0 < y$ then $0 < x \cdot y$.

c) For any x and y in F exactly one of the following holds: $x < y$ or $y < x$ or $x = y$.

The real numbers are often defined as a complete, ordered field that contains the rational numbers. This does not give rise to a unique field and we do not yet know what the word *complete* means. That is the subject of the next chapter.

Much of the work of this course deals not with single real numbers but sets of real numbers. We need the basic vocabulary of Set Theory and Functions.

Definition 2.3 (Sets and Set Operations)**.** Let A and B be sets.

a) A is a *subset* of B, written $A \subset B$, if whenever $x \in A$ then $x \in B$. (Note that we will use $\subset$ where some texts use $\subseteq$.)

b) The *union* of A and B, written $A \cup B$, is the set $A \cup B = \{x \mid x \in A \text{ or } x \in B\}$.

c) The *intersection* of A and B, written $A \cap B$, is the set $A \cap B = \{x \mid x \in A \text{ and } x \in B\}$.

d) The *set difference*, A without B, written $A - B$ or A/B, is the set $A - B = \{x \in A \mid x \notin B\}$. This is also called *the complement of A relative to B*.

e) The *complement* of a set A, with respect to the real numbers $\mathbb{R}$, is the set $A^c = \{x \in \mathbb{R} \mid x \notin A\}$.

f) The *empty set* (or *null set*), $\emptyset$, is the set containing no elements.

g) The *Cartesian Product* of A and B is the set $A \times B = \{(a, b) \mid a \in A \text{ and } b \in B\}$.

Among the most useful sets of real numbers are intervals. Given two real numbers, a and b, with $a < b$, there are four different intervals defined by these two numbers:

- The open interval $(a, b) = \{x \in \mathbb{R} \mid a < x < b\}$.
- The closed interval $[a, b] = \{x \in \mathbb{R} \mid a \leq x \leq b\}$.
- The half-open or half-closed intervals $[a, b) = \{x \in \mathbb{R} \mid a \leq x < b\}$ and $(a, b] = \{x \in \mathbb{R} \mid a < x \leq b\}$.

Using the symbols ∞ and $-\infty$ one defines other intervals:

- $(-\infty, b) = \{x \in \mathbb{R} \mid x < b\}$
- $(-\infty, b] = \{x \in \mathbb{R} \mid x \leq b\}$
- $(a, \infty) = \{x \in \mathbb{R} \mid x > a\}$
- $[a, \infty) = \{x \in \mathbb{R} \mid x \geq a\}$

The entire set of real numbers can be denoted as an interval $\mathbb{R} = (-\infty, \infty)$.
Two special cases need consideration:

- The set $[a, a] = \{a\}$ consists of a alone.
- Finally, any interval of the above form where $a > b$ has no elements and will be named the empty set. For example we write $\emptyset = [2, 1] = \{x \in \mathbb{R} \mid 2 \leq x \leq 1\}$. No real number is simultaneously greater than or equal to 2 and less than or equal to 1.

Definition 2.4 (Functions)**.** Let A and B be sets. A *function* f from A to B, written $f : A \to B$, is a collection f of ordered pairs, (a, b) such that $a \in A$, $b \in B$, and for each a there is exactly one ordered pair (a, b) in the set f. The set A is called the *domain* of f. The set B is called the *co-domain* of f. The *range* of f is the set $\{b \in B \mid (a, b) \in f \text{ for some } a \in A\}$.

Note that the range of f is a subset of the co-domain B but not necessarily equal to B. Also note that a function from A to B is a subset of the Cartesian product of A and B.

Definition 2.5 (Injective Function)**.** Let $f : A \to B$ be a function. We say that f is *one-to-one*, or *injective*, if whenever $a_1 \neq a_2$ for points $a_1, a_2 \in A$ then $f(a_1) \neq f(a_2)$.

Definition 2.6 (Surjective Function)**.** Let $f : A \to B$ be a function. We say that f is *onto*, or *surjective*, if for each $b \in B$ there is at least one $a \in A$ such that $f(a) = b$.

Definition 2.7 (Bijective Function). A function, $f : A \to B$, that is both one-to-one and onto is called a *one-to-one correspondence between A and B* or a *bijection.*

Definition 2.8 (Inverse of a Function). A function $g : B \to A$ is the *inverse of the function* $f : A \to B$ if $g(f(a)) = a$ for every a in A and $f(g(b)) = b$ for every b in B. The inverse of f is typically denoted by f^{-1}, once established as the inverse.

The interactions between sets and functions are very important parts of Real Analysis. We pick out the direct image and the inverse image as especially important.

Definition 2.9 (Direct and Inverse Images). Let $f : A \to B$ be a function.

a) For $U \subset A$, let $f(U) = \{y \in B \mid y = f(x) \text{ for some } x \in U\}$. The set $f(U)$ is called the *direct image of U under f*.

b) For $V \subset B$, let $f^{-1}(V) = \{x \in A \mid f(x) \in V\}$. The set $f^{-1}(V)$ is called the *inverse image of V under f*. Note that the inverse image of V is defined even if f does not have an inverse as a function.

Definition 2.10 (Important Subsets of the Real Numbers). There are several very important subsets of the real numbers which we now define and give notation for.

- $\mathbb{N}$ is the set of natural numbers.
- $\mathbb{Z} = \{\ldots, -2, -1, 0, 1, 2, \ldots\}$ is the set of integers.
- $\mathbb{Q} = \{\frac{p}{q} \mid p \in \mathbb{Z}, q \in \mathbb{N}, \text{ and } \gcd(p, q) = 1\}$ is the set of rational numbers. The rational numbers form an ordered field but not a complete ordered field.
- $\mathbb{R}$ is the set of real numbers. One of the goals of this course is to understand what $\mathbb{R}$ actually is.

Theorem 2.11 (Triangle Inequality). Let x and y be real numbers. Then $|x + y| \leq |x| + |y|$.

Proof. From algebra, $|x + y|^2 = (x + y)^2 = x^2 + 2xy + y^2 = |x|^2 + 2xy + |y|^2$. If x and y are of opposite sign then $2xy < 2|xy|$ and equality holds otherwise. Thus $|x + y|^2 = |x|^2 + 2xy + |y|^2 \leq |x|^2 + 2|x||y| + |y|^2 = (|x| + |y|)^2$. Since the first and last terms are positive we have $|x + y| \leq |x| + |y|$.

Thus we have equality if the numbers have the same sign and strict inequality if they are of different sign. □

Theorem 2.12 (The Cauchy-Schwarz Inequality). Let $\hat{x} = (x_1, x_2, \ldots, x_n)$ and $\hat{y} = (y_1, y_2, \ldots, y_n)$ be n-tuples of real numbers. Then

$$(x_1y_1 + x_2y_2 + \cdots + x_ny_n)^2 \leq (x_1^2 + x_2^2 + \cdots + x_n^2)(y_1^2 + y_2^2 + \cdots + y_n^2).$$

Proof. We will use the notation for dot product to simplify matters. With $\hat{x}$ and $\hat{y}$ as above let $\hat{x} \cdot \hat{y} = x_1y_2 + x_2y_2 + \cdots + x_ny_n$. Note that any n-tuple dotted with itself is a sum of squares and hence is greater than or equal to 0. Consider for a real number λ, the dot product $(\hat{x} + \lambda\hat{y}) \cdot (\hat{x} + \lambda\hat{y}) \geq 0$. Expanded we have $\hat{x} \cdot \hat{x} + 2\lambda\hat{x} \cdot \hat{y} + \lambda^2\hat{y} \cdot \hat{y} \geq 0$. Think of this as a quadratic in λ. Since it is always greater than or equal to 0, the discriminant must be less than or equal to 0. The discriminant is $(2\hat{x} \cdot \hat{y})^2 - 4(\hat{x} \cdot \hat{x})(\hat{y} \cdot \hat{y}) \leq 0$. Dividing by 4 and moving the second term to the right hand side yields $(\hat{x} \cdot \hat{y})^2 \leq (\hat{x} \cdot \hat{x})(\hat{y} \cdot \hat{y})$ or exactly the inequality we seek. □

Theorem 2.13 (The DeMorgan Laws). Let $\{U_\alpha \mid \alpha \in A\}$ be any collection of sets. Then:

(1) $\left(\bigcap_{\alpha\in A} U_\alpha\right)^c = \bigcup_{\alpha\in A} U_\alpha^c$

(2) $\left(\bigcup_{\alpha\in A} U_\alpha\right)^c = \bigcap_{\alpha\in A} U_\alpha^c$

2.1 Exercises

Exercise 2.1. Prove: The sum of a ration number and an irrational number is irrational.

Exercise 2.2. The statement "The product of a rational number and an irrational number is irrational" is not always true. When is the statement false? Rewrite the statement to make it true and prove it.

Exercise 2.3. Find pairs, a and b, of irrational numbers (if possible) fitting the specified condition.

a) $a + b$ is irrational

b) $a + b$ is rational

c) ab is irrational

d) ab is rational

Exercise 2.4. The sum of two rational numbers is always a rational number. It is possible for the sum of two irrational numbers to be rational. For example $\sqrt{7}$ and $7 - \sqrt{7}$. Is it possible for the sum of three irrational numbers to be rational? Either prove that it is impossible or give an example.

Exercise 2.5. Let $x = \frac{1}{\sqrt{3}+\sqrt{7}}$. Find a natural number n such that $0 < \frac{1}{n} < x$.

Exercise 2.6. Find two rational numbers, one greater than $\sqrt{7}$ and one less than $\sqrt{7}$ such that they are each within 0.00001 of $\sqrt{7}$.

Exercise 2.7. Prove: For any sets U, V, and W, $(U \cup V) \cap W = (U \cap W) \cup (V \cap W)$.

Exercise 2.8. Prove: For any sets U and V, $(U \cap V)^c = U^c \cup V^c$ and $(U \cup V)^c = U^c \cap V^c$.

Exercise 2.9. Let $A = B = \mathbb{R}$ (the Real Numbers). Let $f(x) = x^2 - 1$ be a function from A to B. Find each of the following:

a) $f([0, 1])$

b) $f([-1, 2])$

c) $f([1, 2])$

d) $f([0, 2]) \cap f([1, 3])$

e) $f([0, 2]) \cup f([1, 3])$

f) $f^{-1}([0, 3])$

g) $f^{-1}([0, 10])$

h) $f^{-1}([-1, 1])$

Exercise 2.10. Let $f : \mathbb{N} \to \mathbb{N}$ be defined by $f(n) = 2n + 1$. Let E be the set of even natural numbers and T be the set of natural numbers divisible by 3. Find each of the following:

a) $f(E)$

b) $f(T)$

c) $f(E \cup T)$

d) $f(e) \cup f(T)$

e) $f^{-1}(E)$

f) $f^{-1}(T)$

g) $f^{-1}(E \cup T)$

h) $f^{-1}(E) \cup f^{-1}(T)$

Exercise 2.11. Let $f : A \to B$ and let U and V be subsets of A. It is true that $f(U \cup V) = f(U) \cup f(V)$. However, the statement for intersections is not true. Show by example that $f(U \cap V) = f(U) \cap f(V)$ is not necessarily true. Use an example where f is a simple function and the sets are intervals in the real numbers.

Exercise 2.12. Suppose that $f : A \to B$ is a one-to-one function and U and V are subsets of A. Prove that $f(U \cap V) = f(U) \cap f(V)$.

Exercise 2.13. Let $f : A \to B$ and let U and V be subsets of A. Prove that $f^{-1}(U \cup V) = f^{-1}(U) \cup f^{-1}(V)$ and $f^{-1}(U \cap V) = f^{-1}(U) \cap f^{-1}(V)$.

Exercise 2.14. Let $f : A \to B$ and $g : B \to C$ be functions such that $g \circ f : A \to C$ is a bijection (one-to-one and onto). Prove that f is one-to-one and that g is onto. Find an example of two functions that satisfy the statement for which f is *not onto* and g is *not one-to-one*.

Exercise 2.15. Find real numbers x and y such that equality holds in the triangle inequality. Find another pair of numbers x and y such that the equality does not hold. When does equality have to hold?

Exercise 2.16. Negate the statement "For all real numbers a and b, if $a < b$, then there is a natural number n such that $a + \frac{1}{n} < b$." It is not sufficient to write "It is not the case that" followed by the sentence.

Exercise 2.17. Negate the statement "For every $\varepsilon > 0$, there is a $\delta > 0$ such that for all real numbers x if $|x| < \delta$ then $|x^2| < \varepsilon$."

Exercise 2.18. Suppose that a and b are rational numbers. Show that it is impossible that $\sqrt{3} = a + b\sqrt{2}$. You may use the fact that $\sqrt{n}$ is an irrational number if n is a non-square natural number.

Exercise 2.19. Suppose that n, m, and p are natural numbers and $x = \frac{p}{2^n \cdot 5^m}$. Show that x has a terminating decimal expansion.

Exercise 2.20. Suppose that $a_1, a_2, \dots a_n$ are each decimal digits from the set $\{0, 1, 2, \dots, 9\}$. Prove that the repeating decimal expansion given by $x = 0.1_1 a_2 \cdots a_n a_1 a_2 \cdots a_n \cdots$ is a rational number.

Exercise 2.21. Give an example of two functions, f and g, such that $f \neq g$ but $f \circ g = g \circ f$.

Exercise 2.22. Define (if possible) a function $f : [0, 1] \to \mathbb{R}$ such that $f(x) = \frac{1}{x}$ for all $x \in (0, 1]$.

Chapter 3

Moving from $\mathbb{Q}$ to $\mathbb{R}$

One of the most important episodes in the History of Mathematics was the discovery of irrational numbers. The historical record is not clear as to which irrational number was discovered first, to who made the discovery, and to what the method of discovery was. In fact, the discovery was not of an irrational number at all but of a ratio that was not equal to a ratio of natural numbers. The discovery led to the gradual realization that there was a useful set of numbers that contained the rational numbers. Finally in the Nineteenth Century at least two different mathematicians — Dedekind and Cantor — defined the real numbers using the rationals as a starting point. We will adopt a somewhat simpler definition of the real numbers, namely we will conceive of the real numbers, $\mathbb{R}$, as a complete ordered field containing the rational numbers, $\mathbb{Q}$. The word "complete" is the important word here. Field and ordered we have already met. "Complete" roughly means that the real numbers are the rational numbers with all the holes between them filled in. We will make an assumption of "completeness" for the reals. This assumption will be based on a fairly simple idea from set theory, that of the "least upper bound".

Definition 3.1 (Bounds for Sets)**.** A set A of real numbers is said to be *bounded above* if there is a real number, a, such that $x \leq a$ for every $x \in A$. The number a is called an *upper bound* of A. A similar definition holds for *bounded below* and *lower bound.* A set A is *bounded* if it is both bounded above and bounded below.

Definition 3.2 (Least Upper Bound and Greatest Lower Bound)**.** Let A be a set of real numbers. A real number a is called the *least upper bound* (or *supremum*) of A if a is an upper bound of A and if b is any upper bound of A then $b \geq a$. A similar definition holds for the *greatest lower bound* (or *infimum*) of A. We denote the least upper bound of A by $\text{lub}A$ or $\sup A$ ($\text{glb}A$ or $\inf A$ for greatest lower bound).

Example 3.3. What do these first two examples tell you about whether a least upper bound of a set is in the set?

1. $\text{lub}[0, 1] = 1$

2. $\text{lub}(0, 1) = 1$

3. $\text{glb}\left\{1, \frac{1}{2}, \frac{1}{3}, \frac{1}{4}, \ldots\right\} = 0$

Before we explore what least upper bounds might actually be we make an assumption that they exist. This assumption actually defines the real numbers.

Axiom 6 (The Completeness Axiom)**.** A non-empty set of real numbers that is bounded above has a least upper bound.

Theorem 3.4. The empty set does not have a least upper bound.

Proof. Every real number is an upper bound of the empty set and since there is no smallest real number there is no least upper bound for the empty set. How do we know that there is no smallest real number? Suppose x represented the smallest real number. Then $x1$ would be a smaller real number. □

Directly applying the definition of least upper bound is often difficult. We have a criterion for least upper bounds that is a useful alternative. The idea behind this criterion is illustrated in Figure 3.1, below. The real number a is an upper bound of the set A and the criterion claims that a is the least upper bound of A if no matter how close one gets to a from below (closeness determined by $\varepsilon > 0$) there is always an element of A closer to a. In the figure there are several elements of A that are between $a - \varepsilon$ and a. No matter how small $\varepsilon > 0$ is, there will be points of A between $a - \varepsilon$ and a.

Figure 3.1

Theorem 3.5. Let A be a non-empty set and let a be an upper bound of A. Then a is the least upper bound of A, that is $a = \text{lub}A$, if and only if for each $\varepsilon > 0$ there is an $x \in A$ such that $a - \varepsilon < x$.

Proof. For the forward conditional, assume $a = \text{lub}A$ and let $\varepsilon > 0$ be given. If there is no $x \in A$ satisfying $a - \varepsilon < x$ then $a - \varepsilon$ is an upper bound of A that is smaller than its least upper bound, which is a contradiction. Thus the required x must exist.

For the backward conditional: Now assume the condition for upper bound a and suppose that $b = \text{lub}A$ with $b < a$. We need to show that this leads to a contradiction. Let $\varepsilon = \frac{a-b}{2} > 0$. By our assumption there is an $x \in A$ satisfying $a - \varepsilon < x$ but $x > a - \varepsilon = a - \frac{a-b}{2} = \frac{a+b}{2} > b$. Thus there is an $x \in A$ satisfying $b < x$. Thus b is not even an upper bound of A, much less its least upper bound. □

Corollary 3.6. Let A be a non-empty set and let a be a lower bound of A. Then a is the greatest lower bound of A, that is $a = \text{glb}A$, if and only if for each $\varepsilon > 0$ there is an $x \in A$ such that $x < a + \varepsilon$.

Example 3.7. The least upper bound of the interval $(-\infty, 0)$ is 0. Clearly 0 is an upper bound of the set. Let $\varepsilon > 0$ be arbitrary. Let $x = -\frac{\varepsilon}{2}$. Then $0 - \varepsilon = -\varepsilon < -\frac{\varepsilon}{2} = x$.

How does this argument have to be altered to show that $\text{lub}(-1, 0) = 0$? The problem is that choosing a large $\varepsilon > 0$, say $\varepsilon = 10$, would give us $x = -\frac{\varepsilon}{2} = -5$, which is not in $(-1, 0)$. The solution is a bit awkward but not difficult. Suppose $\varepsilon > 0$ is given. If $\varepsilon \geq 1$, let $x = -\frac{1}{4}$. If $\varepsilon < 1$ then let $x = -\frac{\varepsilon}{2}$.

We use the Completeness Axiom almost immediately to prove some basic results about the real numbers. The first result appears to be obvious but we would have a hard time justifying it without the axiom.

Theorem 3.8. The set, $\mathbb{N}$, of natural numbers is not bounded above.

Proof. Suppose that $\mathbb{N}$ is actually bounded above by a real number x. Since the natural numbers is a non-empty set and we have an upper bound for it, then $\mathbb{N}$ must have a least upper bound. Let $y = \text{lub}\mathbb{N}$ and let $\varepsilon = 1$. Then by Theorem 3.5 there is an element p of $\mathbb{N}$ such that $y - 1 < p$. Hence $y < p + 1$. This contradicts that y is an upper bound of $\mathbb{N}$ since $p + 1$ is a natural number. □

Theorem 3.9 (The Archimedean Principle). If $x > 0$ is a real number, there is a natural number n such that $\frac{1}{n} < x$.

Proof. Let x be any positive real number. Then $\frac{1}{x}$ is also a positive real number. There must be a natural number n that is greater than $\frac{1}{x}$ or else $\frac{1}{x}$ would be an upper bound of the natural numbers. We know that there is no such upper bound. Thus there is a natural number n satisfying $0 < \frac{1}{x} < n$. Inverting yields $\frac{1}{n} < x$. □

The Archimedean Principle quickly gives us a fundamental fact about the relation between the rationals, $\mathbb{Q}$, and the reals, $\mathbb{R}$.

Theorem 3.10 (The Density of the Rational Numbers). There is a rational number between any two distinct real numbers.

Proof. Let a and b be real numbers with $a < b$. We need to find a rational number r that satisfies $a < r < b$. By assumption $b - a > 0$ and by the Archimedean Principle (Theorem 3.9) there is a natural number n satisfying $\frac{1}{n} < b - a$. The idea of the proof is that the rational numbers of the form $\frac{m}{n}$ are evenly spaced in the real numbers, separated by a distance of $\frac{1}{n}$. Since a and b are separated by more than the distance $\frac{1}{n}$ there has to be a rational number with denominator n in between a and b. The trick is to find an appropriate numerator, m.

Let $A = \{p \in \mathbb{Z} \mid b \leq \frac{p}{n}\}$. The set S is a non-empty set of integers that is bounded below. Thus it has a smallest element. Call it m. Then $b \leq \frac{m}{n}$. Suppose that $b = \frac{m}{n}$, then $\frac{m-1}{n} < b$. We claim that $a < \frac{m-1}{n}$, which proves the theorem. Since $\frac{m-1}{n} < b = \frac{m}{n}$ we know that $b - \frac{m-1}{n} \leq \frac{1}{n} < b - a$. Hence $a \leq \frac{m-1}{n}$ by rearranging the inequality. If $b < \frac{m}{n}$ then $\frac{m-1}{n} < b$ by the fact that m is the smallest element of S. Again $b - \frac{m-1}{n} \leq \frac{1}{n} < b - a$ and the theorem is proved. □

The same result holds for irrational numbers. Thus between every two distinct real numbers there is an irrational number. The theorem can be made even stronger. Between every two distinct real numbers there are infinitely many rational numbers.

The next theorem is extremely useful even though its statement seems very detailed and specialized.

Theorem 3.11 (The Nested Intervals Theorem). Let $I_n = [a_n, b_n]$ be a non-empty closed interval ($a_n \leq b_n$) for each natural number n. Further suppose that $I_1 \supset I_2 \supset I_3 \supset \cdots$. Then $\bigcap_{n=1}^{\infty} I_n$ is non-empty. In other words, there is a real number x contained in every one of the intervals.

Proof. If $I_n = [a_n, b_n]$, $I_{n+1} = [a_{n+1}, b_{n+1}]$, and $I_{n+1} \subset I_n$ then $a_n \leq a_{n+1}$ and $b_{n+1} \leq b_n$. Further since all of the intervals are contained in $I_1 = [a_1, b_1]$, it follows that $a_n \leq b_1$ for all n. Let $A = \{a_n \mid n \in \mathbb{N}\}$. Then S is non-empty and bounded above (by b_1). Additionally S has a least upper bound by the Completeness Axiom (Axiom 6). Call it x, that is $x = \text{lub}S$. We claim that x is in every interval, thus proving the theorem. Suppose not. Then there is a natural number n such that $x \notin I_n$ or equivalently $x > b_n$. Let $\varepsilon = x - b_n > 0$. By Theorem 2.12, the Cauchy Schwarz Inequality, there must be an element of S, say a_m, such that $x - \varepsilon < a_m$ but since $x - \varepsilon = b_n$ that implies that $b_n < a_m$. This cannot happen but it takes two cases to show that it cannot. First assume that $n \leq m$. Then $b_m \leq b_n$ and it follows that $b_m < a_m$, which implies that I_m is an empty interval, contrary to our assumption. The other case is essentially the same. Thus we have that $x \in I_n$ for all n and the theorem is proved. $\square$

With this theorem in hand we shall be able to prove some very important facts about the real numbers. But that must wait for a few new ideas to come forth.

3.1 Exercises

Exercise 3.1. Let X be the set of all upper bounds of $(0, 1)$. Express X as an interval.

Exercise 3.2. Which of the following sets are bounded above? Which are bounded below? Which are bounded?

a) $P = \{n \in \mathbb{N} \mid n^2 + 1 \text{ is even}\}$

b) $A = \left\{\frac{m}{m+n} \mid m, n \in \mathbb{N}\right\}$

c) $B = \{\frac{m+n}{n} \mid m, n \in \mathbb{N}\}$

Exercise 3.3. Prove that 2 is the least upper bound of the open interval $(-\infty, 2)$.

Exercise 3.4. Prove that 3 is the least upper bound of the open interval $(1, 3)$.

Exercise 3.5. Let A be a set of real numbers. Prove: If $a \in A$ is an upper bound of A then $a = \sup A$.

Exercise 3.6. Find lub$A$ and glb$A$ for each of the following sets.

a) $A = \{n \in \mathbb{N} \mid n^2 + 2 < 50\}$

b) $A = \left\{\frac{m}{m+n} \mid m, n \in \mathbb{N}\right\}$

c) $A = \{\frac{m}{n} \mid m, n \in \mathbb{N} \text{ and } m + n = 15\}$

d) $A = \{\frac{1}{2}, \frac{2}{1}, \frac{2}{3}, \frac{3}{2}\frac{3}{4}, \frac{4}{3}, \dots\}$

e) $A = \left\{1 + \frac{(-1)^n}{n} \mid n \in \mathbb{N}\right\}$

Exercise 3.7. Find a sequence of non-empty open intervals, (a_n, b_n), satisfying $(a_n, b_n) \supseteq (a_{n+1}, b_{n+1})$ such that the intersection of all the intervals is empty.

Exercise 3.8. Find a rational number between $\frac{1}{\sqrt{10}}$ and $\frac{1}{\sqrt{11}}$ and one between $\frac{1}{2+\sqrt{10}}$ and $\frac{1}{2+\sqrt{11}}$.

Exercise 3.9. Let A be a set of real numbers with a least upper bound and let $B = \{-x \mid x \in A\}$. Prove that glb$B$ exists and satisfies glb$B = -$lub$A$.

Exercise 3.10. Consider the statement: "There is a smallest positive real number." Is it true or false, and why?

Exercise 3.11. Give an example of a set of rational numbers that is bounded above but whose least upper bound is not a rational number.

Exercise 3.12. Suppose that lub$A = a$ and lub$B = b$ for non-empty sets of real numbers A and B. Do lub$(A \cup B)$ and lub$(A \cap B)$ always exist? If not, does either sometimes exist and sometimes not? If they do exist, what are they?

Exercise 3.13. Suppose that A and B are non-empty sets that are bounded above. Let $A + B = \{a + b \mid a \in A, b \in B\}$. Prove that lub$(A + B) =$ lub$A +$ lub$B$.

Exercise 3.14. Let $a < b$ be irrational numbers. Prove that there is a rational number p such that $a < p < b$.

Exercise 3.15. Let $I_n = \left[1 - \frac{1}{n}, 1 + \frac{3}{n+1}\right]$ for each natural number n. Find $\bigcap_{n=1}^{\infty} I_n$.

Exercise 3.16. Let $0 < a < 1$. Prove that $a < \sqrt{a}$.

Exercise 3.17. Let $g(x) = \begin{cases} 1 & 0 \leq x \leq 1 \\ 2 & 1 < x < 2 \\ 3 & 2 \leq x \leq 3 \end{cases}$. Suppose that $f : [0,3] \to \mathbb{R}$ satisfies $f(x) \leq g(x)$ for all $x \in [03]$.

a) Is f bounded on $[0,3]$? Justify your answer.

b) Does $\text{lub}\{f(x) \mid x \in [0,3]\}$ exist? Again, justify your answer.

c) Can you compute $\text{lub}\{f(x) \mid x \in [0,3]\}$?

Chapter 4

Cardinality

Although mathematicians had been using the real numbers for thousands of years it was only in the 1800s that Georg Cantor discovered a very important fact about the set of real numbers. He looked carefully at the notion of counting and developed a definition for the number of elements in a set, or the cardinality of the set. He found that the set of rational numbers and the set of irrational numbers, although both infinite, had a different infinity of elements. It would take mathematicians fifty years to sort out the consequences of this difference. Along the way he also essentially invented set theory. His discovery is essential for an understanding of the real numbers.

The definition of cardinality is an abstract definition derived from the process of counting out the elements of a set.

Definition 4.1 (Same Cardinality)**.** Two sets, A and B, have the same cardinality if there is a one-to-one correspondence between them. In this case we write $|A| = |B|$.

When A and B are both finite, we interpret this to mean that A and B have the same number of elements. When A and B are both infinite, they both have infinitely many elements but may not have the same cardinality.

Definition 4.2 (Cardinalities of Sets)**.** A set of real numbers (or actually any set) is *finite* if it is empty or if there is a one-to-one correspondence between it and a set of the form $\{1, 2, 3, \ldots, n\}$ for some natural number n. A set is *infinite* if it is not finite. A set is *countably infinite* if there is a one-to-one correspondence between the set and the set of natural numbers, $\mathbb{N}$. A set is *countable* if it is finite or countably infinite. A set is *uncountable* if it is not countable.

Example 4.3 (Countably Infinite Sets)**.** The following sets are all countably infinite. We provide the bijections between these sets and the set of natural numbers.

1. The integers — $\mathbb{Z}$

   Define $f : \mathbb{N} \to \mathbb{Z}$ by $f(n) = \begin{cases} \frac{n}{2} & \text{if } n \text{ is even} \\ \frac{1-n}{2} & \text{if } n \text{ is odd} \end{cases}$. It is not hard to show that f is a bijection. Thus $|\mathbb{Z}| = |\mathbb{N}|$.

2. The prime numbers.

   Let P be the set of prime numbers. We send natural number n to the nth prime.

3. The even natural numbers — E

Define $f : \mathbb{N} \to E$ by $f(n) = 2n$. This is clearly a bijection and hence $|E| = |\mathbb{N}|$.

4. The rational numbers — $\mathbb{Q}$

   This is somewhat surprising and the full proof is rather detailed. A theorem that we offer without proof helps a great deal in showing that the set of rationals is a countably infinite set.

Theorem 4.4 (The Schroeder-Bernstein Theorem)**.** Suppose that A and B are sets and that we have two one-to-one functions, $f : A \to B$ and $g : B \to A$. Then $|A| = |B|$.

The proof of this theorem is at once elementary (it uses nothing very advanced), interesting, and complicated.

Remark 4.5. It is useful to compare the sizes of sets. From the Schroeder-Bernstein Theorem above we see that two sets have the same number of elements if there are one-to-one functions from each to the other. If we have two sets, A and B, and there is a one-to-one function from A to B, then the sets might have the same number of elements or B might actually have a larger number of elements. We will say that $|A| \leq |B|$ if there is a one-to-one function from A to B. We can then say that A has fewer elements than B and write $|A| < |B|$ if $|A| \leq |B|$ but $|A| \neq |B|$. For example $|\{a, b, c\}| < |\{d, e, f, g, h\}|$.

Countably infinite sets are often called denumerable or enumerable because they can be numbered. That is, if A is a countably infinite set then $A = \{a_1, a_2, a_3, \dots\}$ where each element of A appears in the list exactly once. This follows since if A is countably infinite then there is a bijection $f : \mathbb{N} \to A$ and we define $a_n = f(n)$ for each $n \in \mathbb{N}$. We will often use this characterization of countably infinite sets.

On the way to prove that the rational numbers is a countably infinite set we prove the following:

Theorem 4.6. Let U and V be countably infinite sets. Then $U \times V$ is countably infinite.

Proof. By Theorem 4.4 it suffices to define two one-to-one functions, one from $U \to U \times V$ and one from $U \times V \to U$. The first is quite easy. Send u_k to (u_k, v_1) for each k. For the second function we send (u_m, v_n) to u_k where $k = T_{m+n-1} - m + 1$ is the nth triangular number, that is $T_n = 1 + 2 + \cdots + n = \frac{n(n+1)}{2}$. The first few applications of this function are:

$$
\begin{aligned}
(u_1, v_1) &\to u_1, \\
(u_2, v_1) &\to u_2, \\
(u_1, v_2) &\to u_3, \\
(u_3, v_1) &\to u_4, \\
(u_2, v_2) &\to u_5.
\end{aligned}
$$

We leave the proof that this function is one-to-one to the reader. □

Theorem 4.7. The set $\mathbb{Q}$ is countably infinite.

Proof. A complete proof is full of pesky details. We give a sketch of a proof. By the previous theorem the set $\mathbb{Z} \times \mathbb{N} = \{(p, q) \mid p \in \mathbb{Z}, q \in \mathbb{N}\}$ is a countably infinite set. There is a simple one-to-one function $\mathbb{Q} \to \mathbb{Z} \times \mathbb{N}$ given by $\frac{p}{q} \to (p, q)$ where we take the fraction to be in lowest terms and having a positive denominator. □

One of the major mathematical surprises of the Nineteenth Century was the discovery by Georg Cantor that the real numbers was not a countably infinite set. There are several different proofs but we will use the Nested Interval Theorem (Theorem 3.11) here.

Theorem 4.8. The set $\mathbb{R}$ is uncountable.

Proof. Assume that $\mathbb{R}$ is countably infinite, hence we can write $\mathbb{R} = \{x_1, x_2, x_3, \ldots\}$ where every real number occurs exactly once in the list. Let n be the smallest natural number greater than x_1. Then let $I_1 = [n, n+1]$, a non-empty closed interval. Note that $x_1 \notin I_1$. We break the interval I_1 into three pieces of equal length: $\left[n, n+\frac{1}{3}\right]$, $\left[n+\frac{1}{3}, n+\frac{2}{3}\right]$, and $\left[n+\frac{2}{3}, n+1\right]$. At least one of these intervals does not contain x_2. Let the leftmost one of these be I_2. Similarly divide I_2 into three equal pieces and let I_3 be the leftmost one that does not contain x_3. We proceed to define a non-empty closed interval, I_k, that does not contain x_k. Note that $I_1 \supset I_2 \supset I_3 \supset \cdots$. By the Nested Interval Theorem (Theorem 3.11), there is a real number y that is in every one of the intervals. But by our assumption that $\mathbb{R}$ is countably infinite we know that $y = x_k$ for exactly one natural number k. But then $y = x_k \notin I_k$, which is a contradiction. □

We will conclude with a result of Cantors that introduces the *power set* of a set.

Definition 4.9 (Power Set)**.** Let A be a set. The *power set of* A, $\mathcal{P}(A)$, is the set of all subsets of the set A.

Example 4.10.

1. Let $A = \{a, b, c\}$. Then $\mathcal{P}(A) = \{\emptyset, \{a\}, \{b\}, \{c\}, \{a, b\}, \{a, c\}, \{b, c\}, A\}$. The set A has 3 elements while $\mathcal{P}(A)$ has $2^3 = 8$ elements.

2. The set $\mathcal{P}(\mathbb{N})$ is the set of all possible sets of natural numbers, both finite and infinite. It turns out that $|\mathcal{P}(\mathbb{N})| = |\mathbb{R}|$; that is, the number of subsets of the natural numbers is the same as the number of real numbers.

The next theorem shows us that there are infinitely many different infinities.

Theorem 4.11. Let A be a set. Then there is no onto function from $A \to \mathcal{P}(A)$. In other words, $|A| < |\mathcal{P}(A)|$.

Proof. Suppose that there is an onto function from A to its power set, $\mathcal{P}(A)$. Call it f. If $x \in A$ then $f(x)$ is a subset of A. We define the subset X of A as follows:

$$X = \{x \in A \mid x \notin f(x)\}.$$

Since $f(x)$ is a subset of A it makes sense to ask whether x is an element of that subset. Since f is onto there must be a $y \in A$ satisfying $X = f(y)$. Is $y \in X$?

Assume that $y \in X = \{x \in A \mid x \notin f(x)\}$. By the definition of X, it follows that $y \notin f(y) = X$, which is a contradiction.

Thus we must assume that $y \notin X = f(y)$. But that is precisely the condition that says $y \in X$. Thus if y is in the set, then it is not in the set, and vice-versa. Thus no such onto function can exist. □

By continually forming power sets, starting with the natural numbers, one produces an infinite sequence of ever larger infinite sets:

$$|\mathbb{N}| < |\mathcal{P}(\mathbb{N})| < |\mathcal{P}(\mathcal{P}(\mathbb{N}))| < \cdots.$$

4.1 Exercises

Exercise 4.1. Find an explicit one-to-one and onto function from $3\mathbb{Z} = \{\cdots, -6, -3, 0, 3, 6, \cdots\}$ to $\mathbb{N}$. That means find a formula for such a function and show that it is both one-to-one and onto.

Exercise 4.2. Let $U = \{x_1, x_2, \ldots\}$ and $V = \{y_1, y_2, \ldots\}$ be disjoint, countably infinite sets. Prove that $U \cup V$ is countably infinite. Is the statement still true if the word "disjoint" is removed? Why or why not?

Exercise 4.3. Show that the function which sends (u_m, v_n) to u_k where $k = T_{m+n-1} + m - 1$ in Theorem 4.6 is actually one-to-one.

Exercise 4.4. One can prove that the real numbers are uncountable using what is called "the diagonal argument". Suppose we apply the same argument to the rational numbers lying between 0 and 1. Call that set S, so that $S = \{x \in \mathbb{Q} \mid 0 < x < 1\}$.

> Now assume we have an onto function from $\mathbb{N}$ to S, in other words a listing of all the rational numbers in S. As in the diagonal argument let $S = \{q_1, q_2, \ldots\}$ and let each element of S be expressed as a decimal expansion and if the decimal expansion is terminating change it to a decimal expansion which ends with an infinite sequence of 9s. Now change the kth decimal place of q_k by adding 1 if the digit is less than 9 and making it 8 if it is 9. Call the new digit y_k. Now form the number $0.y_1y_2y_3\cdots$. By the construction this number is not in the list of S. Thus no such onto function exists.

This should show that the rationals are uncountable. What is wrong with this argument?

Exercise 4.5. Let $\mathcal{X} = \{x \in (0,1) \mid x$ has a decimal expansion of 1s, 3s, 5s$\}$. Prove that $\mathcal{X}$ is uncountable.

Exercise 4.6. Let $\mathcal{Y}$ denote the set of numbers in $(0,1)$ with a decimal expansion that contains only 0s and 1s, and only finitely many 0s. Decide if you think $\mathcal{Y}$ is countably infinite or uncountable. Then prove that your decision is the correct one.

Exercise 4.7. Let $f : [0,\infty) \to \mathbb{R}$ be defined by $f(x) = x^2$. Prove in detail that f is one-to-one.

Exercise 4.8. Let $A = \{n \in \mathbb{N} \mid 10 < n^2 + 2 < 200\}$. Find $|A|$, the cardinality of A.

Exercise 4.9. Let each $x \in [0,1]$ be represented as $x = 0.x_1x_2x_3\cdots$, its decimal expansion, where rational numbers which have terminating expansions, other than 0, are given the representation that ends in infinitely many consecutive 9s. Thus $1 = 0.99999\cdots$ and $\frac{1}{4} = 0.249999\cdots$. Define $f : [0,1] \to [0,1] \times [0,1]$ by $f(x) = ((x_1, x_3, x_5, \ldots), (x_2, x_4, x_6, \ldots))$. Prove that f is one-to-one and onto and hence that $[0,1]$ and $[0,1] \times [0,1]$ have the same cardinality.

Chapter 5

Sequences and Series

5.1 Sequences

A very useful object for the study of the real numbers is the sequence. We think of a sequence as a set of real numbers indexed by the natural numbers. Sequences allow us to use countable sets to explore uncountable sets and functions on uncountable sets.

Example 5.1 (Sequences).

1. $\{1, \frac{1}{2}, \frac{1}{3}, \dots, \frac{1}{n}, \dots\}$, $\{a_n = \frac{1}{n} \text{ for all } n \in \mathbb{N}\}$
2. $\{1, 2, 4, 8, \dots, 2^{n-1}, \dots\}$, $\{a_n = 2^{n-1} \text{ for all } n \in \mathbb{N}\}$
3. $\{-1, 1, -1, 1, -1, 1, \dots\}$, $\{a_n = (-1)^n \text{ for all } n \in \mathbb{N}\}$
4. $\{1, \frac{1}{2}, \frac{2}{3}, \frac{3}{5}, \frac{5}{8}, \dots\}$, $\{a_n = ?\}$
5. $\{3, 3.1, 3.14, 3.141, 3.1415, \dots\}$, $\{a_n = ?\}$

We generally define sequences in one of two ways, either by a functional formula or by a recursive formula. Examples 1., 2., and 3. are all given by function formulas. Example 4. is given in a different fashion. Let $a_1 = 1$. Then we define each successive term by applying a formula to the preceding element of the sequence as follows: $a_{n+1} = \frac{1}{1+a_n}$. Thus $a_2 = \frac{1}{1+a_1} = \frac{1}{1+1} = \frac{1}{2}$ and $a_3 = \frac{1}{1+a_2} = \frac{1}{1+\frac{1}{2}} = \frac{2}{3}$. We can give a functional formula for this sequence but it involves the sequence of Fibonacci numbers, F_n, namely $a_n = \frac{F_n}{F_{n+1}}$. More on this later. The fifth sequence comes from the decimal expansion of π.

Now we give the definition of a sequence.

Definition 5.2 (A Sequence of Real Numbers). A *sequence of real numbers* is a function, f, from the natural numbers to the real numbers. That is, $f : \mathbb{N} \to \mathbb{R}$.

How does this definition fit with our examples? If we let $a_n = f(n)$ we have our examples, sometimes with a very simple f and sometimes a complicated one.

We usually denote a sequence in one of the following ways:

$$\{a_n \mid n \in \mathbb{N}\}, \quad \{a_n\}, \quad \{f(n)\}, \quad \text{or} \quad \{a_n = f(n)\}.$$

Often it is convenient to change the indexing slightly and start a sequence with a zeroth term.

$$\{a_0, a_1, a_2, \dots\}$$

The most important question to answer about a sequence is whether or not it converges. The following defines what is meant by the convergence of a sequence. A sequence converges if the sequence has a limit. Sequential limits are the simplest limits we will encounter in Real Analysis.

Definition 5.3 (Sequence Convergence). Let $\{a_n \mid n \in \mathbb{N}\}$ be a sequence of real numbers. We say that $\{a_n\}$ converges to real number L or that the limit of a_n as n approaches infinity is L, written $\lim_{n\to\infty} a_n = L$, if for each $\varepsilon > 0$ there is a natural number N (depending on ε) such that if $n \geq N$ then $|a_n - L| < \varepsilon$.

Proposition 5.4. If $\lim_{n\to\infty} a_n = L$ then $\lim_{n\to\infty} a_{n+1} = L$.

Proof. Let $\varepsilon > 0$ be given. Since $\lim_{n\to\infty} a_n = L$ there is a natural number N such that if $n \geq N$ then $|a_n - L| < \varepsilon$. The nth term of the sequence $\{a_{n+1}\}$ is a_{n+1}, hence if $n \geq N$ then $n+1 \geq N$. Thus $|a_{n+1} - L| < \varepsilon$. □

Example 5.5.

1. $\lim_{n\to\infty} \frac{1}{n} = 0$

Note that the sequence in this example is $\{a_n\} = \{\frac{1}{n}\} = \{1, \frac{1}{2}, \frac{1}{3}, \dots\}$. Now suppose that $\varepsilon > 0$ is an arbitrary real number. Our task is to find a natural number, N, which realizes the definition of convergence. The value of N will depend on ε.

We need to satisfy the following inequality for all $n \geq N$:

$$|a_n - L| = \left|\frac{1}{n} - 0\right| = \left|\frac{1}{n}\right| = \frac{1}{n} < \varepsilon.$$

This is equivalent to $\frac{1}{\varepsilon} < n$. Thus we choose N to be any natural number satisfying $N > \frac{1}{\varepsilon}$. If $n \geq N$ then $n \geq N > \frac{1}{\varepsilon}$, or $\left|\frac{1}{n} - 0\right| = \frac{1}{n} < \varepsilon$.

Let's see how this works if we had a value for ε in mind. Suppose $\varepsilon = 0.14$. We need to choose a natural number N satisfying $N > \frac{1}{0.14} = 7.142857\cdots$. Let $N = 10$. (We might have chosen 8, 9, or 12345.) Then if $n \geq N$ it follows that $\frac{1}{n} < 0.14$. Check it out:

$$a_{11} = \frac{1}{11} = 0.090909\cdots,$$
$$a_{12} = \frac{1}{12} = 0.083333\cdots,$$
$$a_{13} = \frac{1}{13} = 0.076923\cdots, \text{ and so on.}$$

For each $n > 10$, it is indeed true that $|a_n - L| = \frac{1}{n} < 0.14 = \varepsilon$. We've developed a formula so that when an ε is chosen, we know how to calculate N.

2. $\lim_{n\to\infty} \frac{2n^2+3}{n^2-2} = 2$

We first make a computation to determine N in terms of ε.

$$\left|\frac{2n^2+3}{n^2-2} - 2\right| = \left|\frac{2n^2+3}{n^2-2} - \frac{2(n^2-2)}{n^2-2}\right| = \left|\frac{7}{n^2-2}\right| = \frac{7}{n^2-2}$$

The above calculations are true except if $n = 1$, when $\left|\frac{2n^2+3}{n^2-2} - 2\right|$ equals 7. Our task is to make this quantity less than ε.

$$\frac{7}{n^2-2} < \varepsilon$$
$$\frac{7}{\varepsilon} < n^2 - 2$$
$$2 + \frac{7}{\varepsilon} < n^2$$

Finally, we need $n > \sqrt{2 + \frac{7}{\varepsilon}}$. Thus choose N to be any natural number satisfying $N > \sqrt{2 + \frac{7}{\varepsilon}}$.

Proof. Let $\varepsilon > 0$ be given. Choose N to be a natural number such that $N > \sqrt{2 + \frac{7}{\varepsilon}}$. If n is a natural number satisfying $n \geq N$, then $n > \sqrt{2 + \frac{7}{\varepsilon}}$ or equivalently (by our computation) $\left|\frac{2n^2+3}{n^2-2} - 2\right| = \left|\frac{7}{n^2-2}\right| < \varepsilon$. □

Remark 5.6. What does this definition "mean"? One way to look at the definition is to see that it deals with parts of sequences, namely the part of a sequence that leaves out some initial terms. Suppose that $\{a_n\} = \{a_1, a_2, a_3, \ldots\}$ is a sequence and $N = 10$. The "infinite tail" of the given sequence represented by $\{a_N, a_{N+1}, \ldots\} = \{a_{10}, a_{11}, a_{12}, \ldots\}$ simply leaves out the first 9 terms $\{a_1, a_2, \ldots, a_9\}$. Our definition says that given a positive number ε there is an infinite tail of the sequence all of whose terms are contained in the set $(L - \varepsilon, L + \varepsilon)$. Thus if N is chosen for a given ε then we know that $\{a_N, a_{N+1}, a_{N+2}, \ldots\} \subset (L - \varepsilon, L + \varepsilon)$. All but the first $N1$ terms of the sequence are within ε of the limit L. This works no matter how small ε might be (that is if the limit exists).

Suppose that for our second example we are given $\varepsilon = 0.002$. We choose N satisfying

$$N > \sqrt{2 + \frac{7}{0.002}} = \sqrt{2 + 3500} = 59.1777\cdots.$$

Let $N = 70$. Then we know that

$$\{a_{70}, a_{71}, a_{72}, \ldots\} = \{2.00143\cdots, 2.00139\cdots, 2.00135\cdots, \ldots\} \subset (L - \varepsilon, L + \varepsilon) = (1.998, 2.002).$$

We come to a theorem that seems obvious. However, in our definition of limit of a sequence the equals sign is just one symbol in the denotation $\lim_{n\to\infty} a_n = L$, not necessarily an actual equal sign. The following theorem proves that it is an equal sign.

Theorem 5.7 (Uniqueness of Limits). Suppose that $\lim_{n\to\infty} a_n = L$ and $\lim_{n\to\infty} a_n = M$. Then $L = M$.

Proof. If the sequence converges, it cannot simultaneously be getting arbitrarily closer and closer to two different numbers.

Let us assume that both limits hold but that $L \neq M$. Let $\varepsilon = \frac{|L-M|}{2} > 0$. Since $\lim_{n\to\infty} a_n = L$ there is a natural number N_1 such that if $n \geq N_1$ then $|a_n - L| < \varepsilon$. Similarly there is a natural number N_2 such that if $n \geq N_2$ then $|a_n - M| < \varepsilon$. Let N be the larger of N_1 and N_2. By the triangle inequality we have the following:

$$2\varepsilon = |L - M| = |(L - a_n) - (M - a_n)| \leq |a_n - L| + |a_n - M| < \varepsilon + \varepsilon = 2\varepsilon.$$

Thus we have $\varepsilon < \varepsilon$ but no number can be less than itself. Thus we contradict the assumption that $L \neq M$. □

An alternate way to prove the theorem is to take the sequence element, a_N, chosen as above and note that it is within ε of both L and M which are 2ε apart, which is impossible.

Definition 5.8 (Boundedness). A sequence, $\{a_n\}$, is said to be *bounded above* if there is a real number M satisfying $a_n \leq M$ for all $n \in \mathbb{N}$. The number M is called *an upper bound* of the sequence. A sequence, $\{a_n\}$, is said to be *bounded below* if there is a real number L satisfying $L \geq a_n$ for all $n \in \mathbb{N}$. A sequence is *bounded* if it is bounded above and bounded below. This quickly translates to the fact that a sequence is bounded if all its terms are contained in a closed interval, $[L, M]$, for a lower bound L and upper bound M.

Theorem 5.9. If a sequence converges, then it is bounded.

Proof. Suppose that $\lim_{n\to\infty} a_n = L$. We need to show that the sequence is bounded. Consider the case in which $\varepsilon = 1$. There is a natural number, N, such that if $n \geq N$ then $|a_n - L| < 1$. This is equivalent to the inequalities $L - 1 < a_n < L + 1$. Thus the part of the sequence starting at the Nth term is bounded below by $L1$, and bounded above by $L + 1$. Among the terms $\{a_1, a_2, \ldots, a_{N-1}\}$, there may be some that do not lie within these bounds.

Let $A = \min\{a_1, a_2, \ldots, a_{N-1}, L - 1\}$ and $B = \max\{a_1, a_2, \ldots, a_{N-1}, L + 1\}$. Then A is a lower bound for the sequence and B is an upper bound. □

This theorem serves as a quick criterion for the non-convergence of some sequences. If a sequence is unbounded then it is divergent. Thus $\{0, 1, 0, 2, 0, 3, 0, 4, \ldots\}$ is divergent. Boundedness does not imply convergence, however. A very useful sequence is $\{(-1)^n\} = \{-1, 1, -1, 1, -1, 1, \ldots\}$ which is a bounded sequence but not convergent.

Actual computation of limits of sequences is often important. The following theorem is a valuable tool in these computations.

Theorem 5.10 (Algebraic Combinations of Limits). Suppose that $\lim_{n\to\infty} a_n = L$, $\lim_{n\to\infty} b_n = M$, and c is a real number. Then:

(1) $\lim_{n\to\infty}(a_n \pm b_n) = L \pm M$

(2) $\lim_{n\to\infty} ca_n = cL$

(3) $\lim_{n\to\infty} a_n b_n = LM$

(4) If $M \neq 0$ then $\lim_{n\to\infty} \frac{a_n}{b_n} = \frac{L}{M}$.

Proof. We will prove (1) and (3). Part (2) is quite simple to prove while (4) is very messy.

For (1), it does not matter whether we have the plus or minus sign. The proof will be the same. We will assume the plus sign.

Let $\varepsilon > 0$ be given. Since $\lim_{n\to\infty} a_n = L$ there is a natural number N_1 which satisfies the condition that if $n \geq N_1$ then $|a_n - L| < \frac{\varepsilon}{2}$. (We apply the definition for $\frac{\varepsilon}{2}$ rather than ε.) Similarly, since $\lim_{n\to\infty} b_n = M$ there is a natural number N_2 such that if $n \geq N_2$ then $|b_n - M| < \frac{\varepsilon}{2}$. Now let $N = \max\{N_1, N_2\}$. Hence if $n \geq N$ then $n \geq N_1$ and $n \geq N_2$. We use the triangle inequality to finish the proof. Assume that $n \geq N$. Then

$$\begin{aligned}
|(a_n + b_n) - (L + M)| &= |(a_n - L) + (b_n - M)| \\
&\leq |(a_n - L)| + |(b_n - M)| \\
&< \frac{\varepsilon}{2} + \frac{\varepsilon}{2} = \varepsilon.
\end{aligned}$$

For (3): Let $\varepsilon > 0$ be given. Since the sequence $\{b_n\}$ converges, it is bounded. Assume that $|b_n| < K$ for some real number K and all natural numbers, n. We now make some mysterious choices reasons for which will become clear. Since $\{a_n\}$ converges to L, there is a natural number N_1 such that if $n \geq N_1$ then $|a_n - L| < \frac{\varepsilon}{2(K+1)}$. Since $\{b_n\}$ converges to M, there is a natural number N_2 such that if $n \geq N_2$ then $|b_n - M| < \frac{\varepsilon}{2(|L|+1)}$. Now let N be the larger of N_1 and N_2. Assume that $n \geq N$. Then

$$\begin{aligned} |a_n b_n - LM| &= |a_n b_n - b_n L + b_n L - LM| \\ &\leq |a_n b_n - b_n L| + |b_n L - LM| \\ &= |a_n - L||b_n| + |L||b_n - M|. \end{aligned}$$

By the way we have chosen N we have

$$\begin{aligned} |a_n b_n - LM| &\leq |a_n - L||b_n| + |L||b_n - M| \\ &< \left(\frac{\varepsilon}{2(K+1)}\right) K + |L| \left(\frac{\varepsilon}{2(|L|+1)}\right) \\ &< \frac{\varepsilon}{2} + \frac{\varepsilon}{2} = \varepsilon. \end{aligned}$$

The proofs of (1) and (2) are easier and the proof of (4) resembles that of (3) except that it is much more complicated. □

Theorem 5.11. Let $\lim_{n\to\infty} a_n = L$ and $\lim_{n\to\infty} b_n = M$, and let c be a real number.

(1) If $a_n \leq b_n$ for all n greater than some natural number N then $L \leq M$.

(2) If $a_n \leq c$ for all n greater than some natural number N then $L \leq c$.

(3) If $a_n \geq c$ for all n greater than some natural number N then $L \geq c$.

Proof. We actually only need to prove (1) since the others follow from the prior theorem. Suppose that $L > M$ in violation of the theorem. Let $\varepsilon = \frac{L-M}{2} > 0$. Since both sequences converge we can find a natural number K that satisfies the following conditions:

- $K \geq N$,
- for all $k \geq K$, $|a_k - L| < \frac{\varepsilon}{2}$, and
- for all $k \geq K$, $|b_k - M| < \frac{\varepsilon}{2}$.

Thus $|a_K - L| < \frac{\varepsilon}{2}$ and $|b_K - M| < \frac{\varepsilon}{2}$. These inequalities are equivalent to $L - \frac{\varepsilon}{2} < a_K < L + \frac{\varepsilon}{2}$ and $M - \frac{\varepsilon}{2} < b_K < M + \frac{\varepsilon}{2}$. We also have $M = L - 2\varepsilon$. Putting all these together yields:

$$b_K < M + \frac{\varepsilon}{2} = L - 2\varepsilon + \frac{\varepsilon}{2} = L - \frac{3\varepsilon}{2} < L - \frac{\varepsilon}{2} < a_K.$$

This contradicts the assumption that $a_n \leq b_n$ for all n. □

Corollary 5.12. Suppose that $\lim_{n\to\infty} a_n = 0$. Then for a given positive real number M there is a natural number N such that if $n \geq N$ then $|a_n| \leq M$.

Proof. This is a very innocent sounding result but a very useful one nonetheless. Let $\varepsilon = \frac{M}{2} > 0$. Since $\lim_{n\to\infty} a_n = 0$, there is a natural number N such that if $n \geq N$ then

$$|a_n - 0| < \varepsilon = \frac{M}{2} < M.$$ □

The following theorem is known in several different contexts by several different names. For example: The Pinching Lemma, The Squeezing Theorem, The Pinching Theorem, etc.

Theorem 5.13 (The Pinching Lemma)**.** Suppose that the following holds for three given sequences: $a_n \leq b_n \leq c_n$ for all $n \geq K$ for some natural number K. Further suppose that $\lim_{n\to\infty} a_n = \lim_{n\to\infty} c_n = L$. Then $\lim_{n\to\infty} b_n = L$.

Proof. Let $\varepsilon > 0$ be given. By the two given limits there exist natural numbers N_1 and N_2 satisfying the following:

- if $n \geq N_1$ then $|a_n - L| < \frac{\varepsilon}{2}$, and
- if $n \geq N_2$ then $|c_n - L| < \frac{\varepsilon}{2}$,

Let $N = \max\{N_1, N_2, K\}$. Let $n \geq N$. Then $L - \frac{\varepsilon}{2} < a_n \leq b_n \leq c_n < L + \frac{\varepsilon}{2}$, which gives $|b_n - L| < \frac{\varepsilon}{2} < \varepsilon$. Thus $\lim_{n\to\infty} b_n = L$. □

Theorem 5.14. (1) $\lim_{n\to\infty} \frac{1}{n} = 0$

(2) If $|a| < 1$ then $\lim_{n\to\infty} a^n = 0$.

(3) If $a > 0$ then $\lim_{n\to\infty} a^{1/n} = 1$.

The proofs of these important facts will be given later. Actually (1) has already been proven in an example.

We now embark on an important journey through the theory of sequences. We have defined *convergence of a sequence* but our definition of convergence requires that we know the limit of the sequence, namely the real number L in $\lim_{n\to\infty} a_n = L$. In what follows we will consider how to deal with a sequence where no such limit, L, is given and indeed may not exist. This will require several new concepts — a monotone sequence, a subsequence of a given sequence, and two important theorems: The Monotone Convergence Theorem and the Bolzano-Weierstrass Theorem. We end this particular journey with the definition of a Cauchy sequence and prove that Cauchy sequences and convergent sequences are the same.

Definition 5.15 (Monotone Sequences)**.** A sequence $\{a_n\}$ is said to be *increasing* (decreasing) if $a_n \leq a_{n+1}$ $(a_n \geq a_{n+1})$ for all natural numbers n. A sequence is called *monotone* or *monotonic* if it is either an increasing or a decreasing sequence.

Remark 5.16. Sometimes what we have defined is called non-decreasing and increasing is used when the inequality is strict. It turns out that for our purposes the distinction is irrelevant.

It turns out that being monotone is a very powerful property for a sequence. The following theorem displays this power.

Theorem 5.17 (Monotone Convergence Theorem)**.** Suppose that $\{a_n\}$ is an increasing sequence that is bounded above (or decreasing and bounded below). Then $\{a_n\}$ converges.

Proof. Since $\{a_n\}$ is increasing, $a_1 \leq a_2 \leq \cdots \leq a_n \leq a_{n+1} \leq \cdots \leq M$ for some real number M. Then the set of terms of the sequence $\{a_n \mid n \in \mathbb{N}\}$ is a non-empty set that is bounded above. Thus it has a least upper bound. Call that number L. We claim that $\lim_{n\to\infty} a_n = L$.

Let $\varepsilon > 0$ be given. By the criterion for least upper bounds (Theorem 2.12), there is an $x \in \{a_n\}$ such that $L - \varepsilon < x$. Now $x = a_N$ for some natural number N. If $n \geq N$ then $a_N \leq a_n \leq L$. Thus $0 \leq L - a_n < \varepsilon$ or $|a_n - L| < \varepsilon$. Thus $\lim_{n\to\infty} a_n = L$. The proof for the other case uses the criterion for greatest lower bounds. □

Before moving on we will apply this theorem to the proof of Theorem 5.14.(2), that is if $|a| < 1$ then $\lim_{n\to\infty} a^n = 0$.

Proof. Assume $a > 0$. (The case $a = 0$ is trivial.) The sequence $\{a, a^2, a^3, \dots\}$ is a decreasing sequence and is bounded below by 0. Thus it has a limit c. By Proposition 5.4 we have that $\lim_{n\to\infty} a^{n+1} = \lim_{n\to\infty} a^n = c$. By Theorem 5.10, $\lim_{n\to\infty} a^{n+1} = \lim_{n\to\infty} a \cdot a^n = a \lim_{n\to\infty} a^n$. Combining these two results gives us that $ac = c$. Since $a < 1$ it follows that $c = 0$.

Now for the case that $a < 0$. Notice $a = -|a|$. Thus $-|a|^n \leq a^n \leq |a|^n$. Then $\lim_{n\to\infty} |a|^n = 0$ by the first part of our proof. Theorem 5.10 allows for $\lim_{n\to\infty} -|a|^n = -\lim_{n\to\infty} |a|^n = -0 = 0$. By the Pinching Lemma (Theorem 5.13), $\lim_{n\to\infty} a^n = 0$. □

The other crucial new idea is that of a subsequence of a sequence.

Definition 5.18 (Subsequence of a Sequence). Let $\{a_n\}$ be a sequence of real numbers and let $i_1 < i_2 < i_3 < \cdots$ be an increasing sequence of natural numbers. The sequence $\{a_{i_k}\} = \{a_{i_1}, a_{i_2}, \dots\}$ is a *subsequence* of $\{a_n\}$.

Example 5.19.

1. Let $\{a_n\} = \{\frac{1}{n}\} = \{1, \frac{1}{2}, \frac{1}{3}, \frac{1}{4}, \dots\}$. The following are subsequences of this sequence.

$$\left\{\frac{1}{2}, \frac{1}{3}, \frac{1}{4}, \dots\right\} = \{a_{n+1}\}$$
$$\left\{\frac{1}{2}, \frac{1}{4}, \frac{1}{6}, \dots\right\} = \{a_{2n}\}$$
$$\left\{1, \frac{1}{2}, \frac{1}{4}, \frac{1}{8}, \dots\right\} = \{a_{2^{n-1}}\}$$
$$\left\{1, \frac{1}{4}, \frac{1}{9}, \dots\right\} = \{a_{n^2}\}$$

For different reasons the following are not subsequences of this sequence.

$$\left\{1, \frac{1}{2}, \frac{1}{2}, \frac{1}{3}, \dots\right\}, \left\{\frac{1}{2}, 1, \frac{1}{4}, \frac{1}{3}, \dots\right\}, \left\{0, 1, \frac{1}{2}, \dots\right\}$$

2. Let $\{a_n\} = \{(-1)^n\} = \{-1, 1, -1, 1, -1, 1, \dots\}$. Examples of subsequences include the following.

$$\{1, 1, 1, 1, 1, 1, \dots\}$$
$$\{1, -1, 1, -1, 1, -1, \dots\}$$
$$\{-1, 1, -1, 1, 1, -1, 1, 1, 1, -1, 1, 1, 1, 1, \dots\}$$

Theorem 5.20. If $\lim_{n\to\infty} a_n = L$ then $\lim_{n\to\infty} a_{i_n} = L$ for every subsequence of the original sequence.

Proof. Given $\varepsilon > 0$ there is a natural number N such that if $n \geq N$, then $|a_n - L| < \varepsilon$. Let $K = i_N$, the index of the Nth term of the subsequence. Then if $n \geq K = i_N \geq N$ we have $i_N \geq n \geq N$ and $|a_{i_n} - L| < \varepsilon$. □

The next theorem, which turns out to be equivalent to the Completeness Axiom (Axiom 6), will provide us with a candidate for the limit of a sequence when such a limit is unknown.

Theorem 5.21 (The Bolzano-Weierstrass Theorem)**.** Every bounded sequence contains a convergent subsequence.

Proof. Let $\{a_n\}$ be a bounded sequence. In particular suppose that M is a positive real number satisfying $|a_n| \leq M$ for every natural number n. Thus all the terms of the sequence are contained in the closed interval $[-M, M]$. Call this interval I_1. An infinite number of the terms of the given sequence are in either $[-M, 0]$ or $[0, M]$. If not the sequence would have at most finitely many terms and would not even be a sequence. Note that a constant sequence like $\{1, 1, 1, 1, \ldots\}$ has an infinite number of terms, all of them equal to 1. Let I_2 be $[-M, 0]$ if that interval contains infinitely many terms of the sequence and if not let $I_2 = [0, M]$. We proceed to define a sequence $\{I_k\}$ of non-empty closed intervals as follows. Given I_n we divide it into two closed intervals using the midpoint of the interval as an endpoint. One of these intervals (at least one) contains infinitely many terms of the sequence. Let I_{n+1} be the left-most of these intervals. We have thus constructed a nested sequence of non-empty closed intervals:

$$I_1 \supset I_2 \supset \cdots \supset I_n \supset I_{n+1} \supset \cdots.$$

By the Nested Interval Theorem (Theorem 3.11) we know that there is a real number x contained in the intersection of these intervals. Thus $x \in I_n$ for every natural number n. Notice that the length of I_1 is $2M$ and each successive interval has half the length of the preceding one. Thus the length of I_n is $\frac{M}{2^{n-2}}$. Now we choose the appropriate subsequence. Let $a_{i_1} = a_1$. Let a_{i_2} be any term of the sequence lying in interval I_2 with $i_2 > i_1 = 1$. We continue on letting a_{i_k} be any sequence element in I_k with $i_k > i_{k-1}$. This gives us a subsequence of the original sequence which we claim converges to x. Let $\varepsilon > 0$ be given. Since $\lim_{n\to\infty} \frac{M}{2^{n-2}} = 0$, there is a natural number N such that the length of interval I_N is less than ε. The number x is also in this interval. Every subsequential element a_{i_n} is also in I_n if $n \geq N$. Thus if $n \geq N$ we have $|a_{i_n} - x| < \varepsilon$. Thus we have $\lim_{n\to\infty} a_{i_n} = x$. □

We have defined a convergent sequence as one in which all the terms eventually approach a particular number called the limit of the sequence. There is a related notion for a sequence in which the terms approach each other instead of a single number. We will show that the two notions are equivalent. These sequences are called Cauchy sequences.

Definition 5.22 (A Cauchy Sequence). A sequence $\{a_n\}$ is called *Cauchy* or *a Cauchy Sequence* if for each $\varepsilon > 0$ there is a natural number N such that if $m, n \geq N$ then $|a_m - a_n| < \varepsilon$.

It is useful to compare the wording of the definition of a convergent sequence (Definition 5.3) and that of a Cauchy sequence (Definition 5.22). What are the differences in the two definitions?

Theorem 5.23. A sequence is Cauchy if and only if it is convergent.

Proof. First we prove that a convergent sequence is Cauchy. Suppose that $\lim_{n\to\infty} a_n = L$. Let $\varepsilon > 0$ be given. Then there is a natural number N such that if $n \geq N$ then $|a_n - L| < \frac{\varepsilon}{2}$. Now assume that $m, n \geq N$. Then

$$|a_m - a_n| = |a_m - L + L - a_n| \leq |a_m - L| + |a_n - L| < \frac{\varepsilon}{2} + \frac{\varepsilon}{2} = \varepsilon.$$

The sequence is Cauchy.

What holds us back from proving the theorem in the other direction is that we dont know that L exists. We know that the terms of the sequence are getting closer to each other but we dont know that there is a number L that they are getting close to. This is where the Bolzano-Weierstrass Theorem (Theorem 5.21) comes in. We now prove that every Cauchy sequence is convergent.

Suppose $\{a_n\}$ is a Cauchy sequence. We prove that it is also bounded so that we can use the previous theorem. Let $\varepsilon = 1$. There is a natural number N such that if $m, n \geq N$ then $|a_m - a_n| < 1$. Let $A = \min\{a_1, a_2, \ldots a_{N-1}, a_N - 1\}$ and $B = \max\{a_1, a_2, \ldots, a_{N-1}, a_N + 1\}$. Then the entire sequence is contained in $[A, B]$, and hence bounded. The original sequence has a convergent subsequence $\{a_{i_n}\}$. Suppose it converges to a. Let $\varepsilon > 0$ be given. There is a natural number K such that if $i_n \geq K$ then $|a_{i_n} - a| < \frac{\varepsilon}{2}$. There is also a natural number L such that if $m, n \geq L$ then $|a_m - a_n| < \frac{\varepsilon}{2}$ since we started with a Cauchy sequence. Now let N be the larger of K and L. Then if $n \geq N$ we have

$$|a_n - a| = |a_n - a_N + a_N - a| \leq |a_n - a_N| + |a_N - a| < \frac{\varepsilon}{2} + \frac{\varepsilon}{2} = \varepsilon.$$

We have convergence. □

The notion of a Cauchy sequence is very important. It was used in one of the first actual definitions of the set of real numbers. A real number was defined (by Georg Cantor) as an equivalence class of Cauchy sequences of rational numbers. For us a more immediate application of Cauchy sequences will be in the development of infinite series, a special form of a sequence.

5.2 Series

Infinite series are particularly useful examples of sequences which we now define. We will introduce several important convergence tests and the notion of absolute convergence.

Definition 5.24 (Infinite Series)**.** Let $\{a_n\}$ be a sequence of real numbers. Let $s_n = a_1 + a_2 + \cdots + a_n = \sum_{k=1}^{n} a_k$. The number s_n is called the *n-th partial sum* of the sequence. The sequence $\{s_n\}$ is called an *infinite series* and denoted by $\sum_{k=1}^{\infty} a_k$. If $\lim_{n\to\infty} s_n$ exists, that is $\{s_n\}$ converges, then we say the infinite series converges. If $\lim_{n\to\infty} s_n = s$, we write $\sum_{k=1}^{\infty} a_k = s$ and call s the sum of the series.

Remark 5.25. It is important to keep in mind that the symbol $\sum_{k=1}^{\infty} a_k$ is simply the name of the sequence $\{s_n\}$ of partial sums. To write $\sum_{k=1}^{\infty} a_k = s$ does not mean that we have added all the terms of the original sequence. We can only add finitely many numbers. The number s represents a limit of finite sums. There are very good reasons for not viewing an infinite series as a sum of infinitely many terms but we will get to that later.

Example 5.26 (A Telescoping Series). Consider $\sum\limits_{k=1}^{\infty} \frac{1}{k^2+k}$. This is the infinite series derived from the sequence $\{a_n\} = \left\{\frac{1}{n^2+n}\right\}$. What is s_n?

The first few terms of $\{a_n\} = \left\{\frac{1}{n^2+n}\right\}$ are $\left\{\frac{1}{2}, \frac{1}{6}, \frac{1}{12}, \frac{1}{20}, \frac{1}{30}, \dots\right\} = \left\{\frac{1}{1\cdot 2}, \frac{1}{2\cdot 3}, \frac{1}{3\cdot 4}, \frac{1}{4\cdot 5}, \dots\right\}$. We compute the first few partial sums:

$$\begin{aligned}
s_1 &= a_1 = \frac{1}{2} \\
s_2 &= a_1 + a_2 = \frac{1}{2} + \frac{1}{6} = \frac{2}{3} \\
s_3 &= a_1 + a_2 + a_3 = s_2 + a_3 = \frac{1}{2} + \frac{1}{6} + \frac{1}{12} = \frac{2}{3} + \frac{1}{12} = \frac{3}{4} \\
s_4 &= s_3 + a_4 = \frac{3}{4} + \frac{1}{20} = \frac{4}{5}
\end{aligned}$$

If we generalize from the pattern (this is not a proof) we would say that $s_n = \frac{n}{n+1}$. If this is true then $\sum\limits_{k=1}^{\infty} \frac{1}{k^2+k} = \lim_{n\to\infty} s_n = \lim_{n\to\infty} \frac{n}{n+1} = 1$. The proof depends on the fact that

$$a_n = \frac{1}{n^2+n} = \frac{1}{n(n+1)} = \frac{1}{n} - \frac{1}{n+1},$$

which is easy to check. Then

$$\begin{aligned}
s_n &= a_1 + \cdots + a_n \\
&= \left(1 - \frac{1}{2}\right) + \left(\frac{1}{2} - \frac{1}{3}\right) + \left(\frac{1}{3} - \frac{1}{4}\right) + \cdots + \left(\frac{1}{n-1} - \frac{1}{n}\right) + \left(\frac{1}{n} - \frac{1}{n+1}\right) \\
&= 1 - \frac{1}{n+1} = \frac{n}{n+1}.
\end{aligned}$$

Our next example and theorem introduce the most important of all infinite series and give a complete description of its convergence.

Example 5.27 (The Geometric Series). In this example our sequence begins with index 0, that is $\{a_n\} = \{a_0, a_1, a_2, \dots\}$. Let x be a real number. Define $a_n = x^n$ for $n = 0, 1, 2, \dots$. We are interested in the infinite series $\sum\limits_{k=0}^{\infty} x^k$, called a *geometric series*. Notice that for a fixed value of x we have $s_n = 1 + x + x^2 + \cdots + x^n$.

If $x = 1$ then $s_n = 1 + 1 + \cdots + 1 = n + 1$. The sequence $\{s_n\}$ of partial sums is just $\{n+1\}$. Clearly $\lim_{s\to\infty} s_n = \lim_{n\to\infty} n + 1$ is not a finite number. Thus the geometric series diverges if $x = 1$.

We now assume that x is different from 1. Then

$$(1-x)s_n = (1-x)(1 + x + x^2 + \cdots + x^n) = 1 - x^{n+1}.$$

Thus we have $s_n = \frac{1-x^{n+1}}{1-x}$, which is defined since $x \neq 1$. Then if it exists,

$$\sum_{k=0}^{\infty} x^k = \lim_{n\to\infty}\left(\frac{1-x^{n+1}}{1-x}\right) = \frac{1-\lim_{n\to\infty} x^{n+1}}{1-x}.$$

Theorem 5.14 tells us that $\lim_{n\to\infty} x^{n+1}0$ if $|x| < 1$. It is easy to see that the limit does not exist if we have $x = -1$ or $|x| > 1$. Thus we have:

$$\sum_{k=0}^{\infty} x^k = \begin{cases} \frac{1}{1-x} & \text{if } |x| < 1 \\ \text{diverges} & \text{if } |x| \geq 1 \end{cases}.$$

Proposition 5.28. For any integer m, $\sum\limits_{k=m}^{\infty} x^k = \begin{cases} \frac{x^m}{1-x} & \text{if } |x| < 1 \\ \text{undefined} & \text{if } |x| \geq 1 \end{cases}.$

Proof. We are given an integer m, which represents the power of x in the leading term of the series. We compute the nth partial sum of the series:

$$s_n = x^m + x^{m+1} + x^{m+2} + \cdots + x^{m+n} = x^m(1 + x + \cdots + x^n).$$

By our work on the geometric series we know that this sequence of partial sums, $\{s_n\}$, diverges if $|x| \geq 1$. The partial sums will converge if $|x| < 1$. What is the limit? As $s_n = x^m(1+x+\cdots+x^n)$, thus the partial sums converge to $\frac{x^m}{1-x}$ in this case. □

Theorem 5.29. Let $\sum\limits_{k=1}^{\infty} a_k$ and $\sum\limits_{k=1}^{\infty} b_k$ be convergent series converging to A and B respectively and let c be a real number. Then:

(1) $\sum\limits_{k=1}^{\infty} (a_k + b_k) = A + B$, and

(2) $\sum\limits_{k=1}^{\infty} ca_k = cA$.

Proof. (1) Let $s_n = a_1 + \cdots + a_n$ and $t_n = b_1 + \cdots b_n$ be the nth partial sums of $\sum\limits_{k=1}^{\infty} a_k$ and $\sum\limits_{k=1}^{\infty} b_k$ respectively. Thus $\lim_{n\to\infty} s_n = A$ and $\lim_{n\to\infty} t_n = B$. Let $u_n = (a_1 + b_1) + \cdots + (a_n + b_n)$ be the nth partial sum of the series $\sum\limits_{k=1}^{\infty} (a_k + b_k)$. Since this is a finite sum we can apply commutativity and associativity to yield $u_n = s_n + t_n$. In other words,

$$(a_1 + a_2 + \cdots + a_n) + (b_1 + b_2 + \cdots + b_n) = (a_1 + b_1) + (a_2 + b_2) + \cdots + (a_n + b_n).$$

Using the limit laws from Theorem 5.10, $A+B = \lim_{n\to\infty} s_n + \lim_{n\to\infty} t_n = \lim_{n\to\infty} u_n = \sum\limits_{k=1}^{\infty} (a_k + b_k)$.

(2) The proof here is accomplished by using distributivity on the partial sums of $\sum\limits_{k=1}^{\infty} ca_k$. □

Remark 5.30. No result like Theorem 5.29 holds for series of the form $\sum_{k=1}^{\infty} a_k b_k$. There are examples for which $\sum_{k=1}^{\infty} a_k$ and $\sum_{k=1}^{\infty} b_k$ both converge while $\sum_{k=1}^{\infty} a_k b_k$ diverges and examples of just the opposite, that is $\sum_{k=1}^{\infty} a_k$ and $\sum_{k=1}^{\infty} b_k$ both diverge while $\sum_{k=1}^{\infty} a_k b_k$ converges.

Theorem 5.31 (The Cauchy Convergence Criterion)**.** Let $\sum_{k=1}^{\infty} a_k$ be an infinite series. Then $\sum_{k=1}^{\infty} a_k$ converges if and only if for each $\varepsilon > 0$ there is a natural number N such that if $n > m \geq N$ then $|a_{m+1} + a_{m+2} + \cdots a_n| < \varepsilon$.

Proof. Let $\{s_n\}$ be the sequence of partial sums of the sequence $\{a_k\}$. Now assume that $\sum_{k=1}^{\infty} a_k$ converges. This is equivalent to the sequence $\{s_n\}$ being convergent. But being convergent and being Cauchy are equivalent (Theorem 5.23). Thus $\{s_n\}$ is also a Cauchy sequence. This is equivalent by definition to the following condition: given $\varepsilon > 0$, there is a natural number N such that if $n > m \geq N$ then $|s_n - s_m| < \varepsilon$. What is $|s_n - s_m|$?

$$|s_n - s_m| = |(a_1 + a_2 + \cdots + a_n) - (a_1 + a_2 + \cdots + a_m)| = |(a_{m+1} + a_{m+2} + \cdots + a_n|$$

Thus $|a_{m+1} + a_{m+2} + \cdots + a_n| < \varepsilon$. Similarly if the series diverges, the sequence of partial sums is not Cauchy and the condition is not satisfied. □

Convergence Tests

An interesting and surprising fact that is reasonably hard to prove is that $\sum_{k=1}^{\infty} \frac{1}{k^2} = \frac{\pi^2}{6}$. The problem of determining to what number this series converges is called "the Basel problem", which was posed originally in the 1600s. Euler solved this problem in the 1700s by establishing the equality described. There are many interesting proofs for this fact with rich interactions of Number Theory and Analysis. Oftentimes, though, the real problem is to determine convergence of a series. Its value will come from some other source. The next several theorems develop convergence tests for series. Some of the tests tell immediately if a series converges or diverges but only rarely does a test say what the sum of the series actually is. Some of the tests only say that a series diverges, and are indeterminate for many series.

Many of the proofs of these tests are left to the exercises at the end of the chapter.

Theorem 5.32 (The Comparison Tests)**.** Suppose that $\{a_n\}$ and $\{b_n\}$ are sequences of non-negative real numbers. Further suppose that for some natural number N if $n \geq N$ then $a_n \leq b_n$. Then:

(1) If $\sum_{k=1}^{\infty} a_k$ diverges then $\sum_{k=1}^{\infty} b_k$ diverges.

(2) If $\sum_{k=1}^{\infty} b_k$ converges then $\sum_{k=1}^{\infty} a_k$ converges.

Proof. For (2): We are assuming that $0 \leq a_n \leq b_n$ for all $n \geq N$. Since $\sum_{k=1}^{\infty} b_k$ converges we can use the Cauchy criterion from Theorem 5.31. Given $\varepsilon > 0$, there is a natural number M satisfying $N \leq M$ and if $n > m \geq M$ then $|b_{m+1}+b_{m+2}+\cdots+b_n| < \varepsilon$. Since all the numbers are non-negative we have $b_{m+1} + b_{m+2} + \cdots b_n = |b_{m+1} + b_{m+2} + \cdots + b_n| < \varepsilon$. By our assumption on the sequences we also have

$$a_{m+1} + a_{m+2} + \cdots + a_n \leq b_{m+1} + b_{m+2} + \cdots b_n.$$

This gives us

$$\begin{aligned} |a_{m+1} + a_{m+2} + \cdots + a_n| &= a_{m+1} + a_{m+2} + \cdots + a_n \\ &\leq b_{m+1} + b_{m+2} + \cdots b_n \\ &= |b_{m+1} + b_{m+2} + \cdots + b_n| \\ &< \varepsilon. \end{aligned}$$

Hence $\sum\limits_{k=1}^{\infty} a_k$ converges by the Cauchy criterion. The statement in (2) is just the contrapositive of the statement in (1) so it is also true. □

Theorem 5.33. If $\sum\limits_{k=1}^{\infty} a_k$ converges, then $\lim_{n\to\infty} a_n = 0$.

Proof. Let $\{s_n\}$ be the sequence of partial sums of the sequence $\{a_k\}$. Since $\sum\limits_{k=1}^{\infty} a_k$ converges so does the sequence $\{s_n\}$. Let $\lim_{n\to\infty} s_n = s$. Then we also have $\lim_{n\to\infty} s_{n+1} = s$ and $\lim_{n\to\infty} s_{n+1} - s = s - s = 0$. But $s_{n+1} - s_n = a_{n+1}$. Thus we have $\lim_{n\to\infty} a_{n+1} = \lim_{n\to\infty} a_n = 0$. □

Note that the converse of this theorem does not hold. Consider the series $\sum\limits_{k=1}^{\infty} \frac{1}{k}$ which diverges but whose terms converge to 0. This theorem is the basis for what is sometimes called the "Negative Test" or the "Divergence Test": *Given an infinite series,* $\sum\limits_{k=1}^{\infty} a_k$, *if* $\lim_{n\to\infty} a_n \neq 0$ *then the series diverges.* This test never proves convergence.

Theorem 5.34 (The p-Test)**.** The series $\sum\limits_{k=1}^{\infty} \frac{1}{k^p}$ converges if $p > 1$ and diverges otherwise. In particular, the Harmonic Series, $\sum\limits_{k=1}^{\infty} \frac{1}{k}$ diverges while $\sum\limits_{k=1}^{\infty} \frac{1}{k^2}$ converges.

Proof. The easiest proof of this theorem uses the Integral Test. Since we have yet to define an integral we will prove this result using the Geometric Series and the Comparison Test (Theorem 5.32).

First we assume that $p > 1$. In the series $\sum\limits_{k=1}^{\infty} \frac{1}{k^p}$ we are going to replace many terms with larger terms. For example we will replace $\frac{1}{3^p}$ with $\frac{1}{2^p}$, which is larger, and $\frac{1}{5^p}$, $\frac{1}{6^p}$, $\frac{1}{7^p}$ each with $\frac{1}{4^p}$, which is again larger. This results in the following:

$$\begin{aligned}\frac{1}{1^p}+\frac{1}{2^p}+\frac{1}{3^p}+\frac{1}{4^p}+\frac{1}{5^p}+\frac{1}{6^p}+\frac{1}{7^p}+\frac{1}{8^p}+\cdots &< \frac{1}{1^p}+\frac{1}{2^p}+\frac{1}{2^p}+\frac{1}{4^p}+\frac{1}{4^p}+\frac{1}{4^p}+\frac{1}{4^p}+\frac{1}{8^p}+\cdots\\ &= \frac{1}{1^p}+\frac{2}{2^p}+\frac{4}{4^p}+\frac{8}{8^p}+\cdots\\ &= 1+\frac{1}{2^{p-1}}+\frac{1}{4^{p-1}}+\frac{1}{8^{p-1}}+\cdots\\ &= 1+\left(\frac{1}{2^{p-1}}\right)+\left(\frac{1}{2^{p-1}}\right)^2+\left(\frac{1}{2^{p-1}}\right)^3+\cdots\\ &= \sum_{k=1}^{\infty}\left(\frac{1}{2^{p-1}}\right)^k.\end{aligned}$$

Since $p > 1$ it follows that $\left(\frac{1}{2^{p-1}}\right) < 1$ and hence the Geometric Series $\sum_{k=1}^{\infty}\left(\frac{1}{2^{p-1}}\right)^k$ converges. The original series thus converges by the Comparison Test (Theorem 5.32).

Next we consider the case $p = 1$. This series is called the Harmonic Series: $\sum_{k=1}^{\infty}\frac{1}{k}$. We will show this series diverges using a similar kind of replacement of terms as in the case for $p > 1$. Instead of rounding 2^k terms of the infinite series up to the nearest $\left(\frac{1}{2^p}\right)^k$, creating intermediate sums of $\left(\frac{1}{2^{p-1}}\right)^{k-1}$ and resulting in a Geometric Series, we'll round 2^{k-1} terms *down* to $\frac{1}{2^k}$ and create intermediate sums of $\frac{1}{2}$. We will consider the partial sums of this series. Let $s_n = \sum_{k=1}^{n}\frac{1}{k} = 1+\frac{1}{2}+\cdots+\frac{1}{n}$.

$$\begin{aligned}s_1 &= 1\\ s_2 &= 1+\frac{1}{2}=\frac{3}{2}\\ s_4 &= 1+\frac{1}{2}+\frac{1}{3}+\frac{1}{4} > 1+\frac{1}{2}+\frac{1}{4}+\frac{1}{4} = 2,\end{aligned}$$

after replacing $\frac{1}{3}$ with the smaller $\frac{1}{4}$. After replacing $\frac{1}{5}, \frac{1}{6}, \frac{1}{7}$ with the smaller $\frac{1}{8}$, we estimate s_8:

$$\begin{aligned}s_8 &= 1+\frac{1}{2}+\frac{1}{3}+\frac{1}{4}+\frac{1}{5}+\frac{1}{6}+\frac{1}{7}+\frac{1}{8}\\ &> 1+\frac{1}{2}+\frac{1}{4}+\frac{1}{4}+\frac{1}{8}+\frac{1}{8}+\frac{1}{8}+\frac{1}{8}\\ &= 1+\frac{1}{2}+\frac{1}{2}+\frac{1}{2}=\frac{5}{2}.\end{aligned}$$

Using induction, one can flesh out the details to prove that $s_{2^n} \geq \frac{n+2}{2}$ for all $n \in \mathbb{N}$. Thus the sequence of partial sums s_n is unbounded and hence the Harmonic Series must diverge.

Assume that $0 < p < 1$. Then for every positive integet n, we see $\frac{1}{n} \leq \frac{1}{n^p}$ and hence the series diverges, again by the Comparison Test (Theorem 5.32). The case $p \leq 0$ is handled very easily since $\lim_{n\to\infty}\frac{1}{n^p} \neq 0$ and hence the series diverges by the contrapositive of Theorem 5.33 (The Negative Test). □

Theorem 5.35 (Alternating Series Test). Let $\{a_n\}$ be a sequence satisfying the following conditions:

(1) $a_1 \geq a_2 \geq a_3 \geq \cdots \geq a_n \geq a_{n+1} \geq \cdots \geq 0$, and

(2) $\lim_{n\to\infty} a_n = 0$.

Then $\sum_{k=1}^{\infty}(-1)^{k+1}a_k = a_1 - a_2 + a_3 - \cdots$ converges.

Proof. We consider two subsequences of the sequence of partial sums. Let $\{s_{2n}\}$ be the partial sums over an even number of elements.

$$s_{2n} = a_1 - a_2 + a_3 - a_4 + \cdots + a_{2n-1} - a_{2n}$$

Note that $s_{2n+2} = s_{2n} + a_{2n+1} - a_{2n+2}$. By condition (1), $a_{2n+1} - a_{2n+2} \geq 0$. Hence $\{s_{2n}\}$ is an increasing sequence. Rewriting s_{2n} as

$$s_{2n} = a_1 - (a_2 - a_3) - (a_4 - a_5) - \cdots - (a_{2n-2} - a_{2n-1}) - a_{2n}$$

and noting that $a_{2k} - a_{2k+1} \geq 0$, we see that $s_{2n} \leq a_1$ for all n. Hence $\{s_{2n}\}$ is increasing and bounded above, hence convergent (Theorem 5.17). Let $\lim_{n\to\infty} s_{2n} = u$.

Now we consider $\{s_{2n-1}\}$, the partial sums over odd numbers of elements.

$$s_{2n-1} = a_1 - a_2 + a_3 - a_4 + \cdots + a_{2n-1}$$

We consider

$$s_{2n+1} = a_1 - a_2 + a_3 - a_4 + \cdots + a_{2n-1} - a_{2n} + a_{2n+1} = s_{2n-1} - (a_{2n} - a_{2n+1}) \leq s_{2n-1}.$$

Thus $\{s_{2n-1}\}$ is a decreasing sequence. Similarly to the other case $a_1 - a_2$ is a lower bound to $\{s_{2n-1}\}$, hence $\{s_{2n-1}\}$ is convergent. Let $\lim_{n\to\infty} s_{2n-1} = v$.

To complete the proof we compute

$$0 = \lim_{n\to\infty} a_{2n} = \lim_{n\to\infty}(s_{2n}s_{2n-1}) = \lim_{n\to\infty} s_{2n} - \lim_{n\to\infty} s_{2n-1} = u - v.$$

Thus $u = v$ and $\{s_n\}$ converges. □

Definition 5.36 (Absolute Convergence). A series $\sum_{k=1}^{\infty} a_k$ is said to *converge absolutely* or be *absolutely convergent* if the series $\sum_{k=1}^{\infty} |a_k|$ converges. If $\sum_{k=1}^{\infty} a_k$ converges but $\sum_{k=1}^{\infty} |a_k|$ diverges we say that $\sum_{k=1}^{\infty} a_k$ *converges conditionally.*

Theorem 5.37. A series that converges absolutely, converges.

Proof. This follows quickly using the Cauchy criterion for convergence of a series. □

Definition 5.38 (Rearrangement of a Series). Let $\sum_{k=1}^{\infty} a_k$ be a series. The series $\sum_{k=1}^{\infty} b_k$ is a *rearrangement* of $\sum_{k=1}^{\infty} a_k$ if there is a one-to-one correspondence $f : \mathbb{N} \to \mathbb{N}$ such that $a_k = b_{f(k)}$ for all k.

Remark 5.39. Consider the alternating series $\sum_{k=1}^{\infty} \frac{(-1)^{k+1}}{k} = 1 - \frac{1}{2} + \frac{1}{3} - \frac{1}{4} + \cdots$. It converges to $\ln(2)$. If the series is rearranged so that two positive terms precede each negative term then we have $1 + \frac{1}{3} - \frac{1}{2} + \frac{1}{5} + \frac{1}{7} - \frac{1}{4} + \frac{1}{9} + \cdots$ which converges to $\frac{3}{2}\ln(2)$. We can actually prove that if a series has two rearrangements that converge to different numbers then for any real number r there is a rearrangement of the series that converges to r. This is the motivation for the next theorem.

Theorem 5.40. If a series converges absolutely then every rearrangement of the series converges to the same value. If a series converges conditionally then for each real number r there is a rearrangement that converges to r.

Theorem 5.41 (Limit Comparison Test). Suppose that $\{a_n\}$ and $\{b_n\}$ are sequences of positive real numbers and that $\lim_{n\to\infty} \frac{a_n}{b_n} = L$ with $0 < L < \infty$. Then both infinite series $\sum_{k=1}^{\infty} a_k$ and $\sum_{k=1}^{\infty} b_k$ converge or both diverge.

For the proof, see Exercise 5.35. Note that when L is equal to 0 or equal to infinity there are deducations that can be made about the convergence of the two series but the conclusions are not as strong as in the theorem.

Theorem 5.42 (Ratio Test). Suppose that $\{a_n\}$ is a sequence of non-zero real numbers and that $\lambda = \lim_{n\to\infty} \left|\frac{a_{n+1}}{a_n}\right|$. Then:

(1) if $\lambda < 1$, then the series $\sum_{k=1}^{\infty} a_k$ converges absolutely;

(2) if $\lambda > 1$, then the series $\sum_{k=1}^{\infty} a_k$ diverges; and

(3) if $\lambda = 1$, then the test gives no information about the convergence of $\sum_{k=1}^{\infty} a_k$.

5.3 Exercises

Exercise 5.1. Prove that $\lim_{n\to\infty}\left(1+\frac{(-1)^n}{n}\right)=1$.

Exercise 5.2. Give an ε — N proof that $\lim_{n\to\infty}\frac{2n-1}{n+4}=2$.

Exercise 5.3. Find $\lim_{n\to\infty}\frac{(n+(-1)^n)}{2n+1}$. Prove your result.

Exercise 5.4. Prove that $\lim_{n\to\infty}\frac{\sin(\frac{1}{n})}{n}=0$.

Exercise 5.5. **Definition**: Let $\{a_n\}$ be a sequence of real numbers and L a real number. We say that $\text{Cbd}_{n\to\infty}a_n = L$ if for each natural number N there is an $\varepsilon > 0$ such that if $n \geq N$, then $|a_n - L| < \varepsilon$.

Prove or disprove that $\text{Cbd}_{n\to\infty}(-1)^n = 0$.

Exercise 5.6. Let the sequence $\{a_n\}$ be defined by 1.) $a_1 = 1$ and 2.) For all $n \geq 1$, $a_{n+1} = \frac{a_n}{3}+\frac{1}{4}$. Prove by induction that the sequence is bounded below and is decreasing. Does $\lim_{n\to\infty} a_n$ exist? If so, what is the limit?

Exercise 5.7. Suppose that $a_1 = 0$ in Exercise 5.6. How should that problem be changed?

Exercise 5.8. Let $a_1 = 1$ and $a_{n+1} = \frac{2a_n}{5}+\frac{4}{5}$ for each natural number $n \geq 2$.

a) Find the first four terms of the sequence.

b) Does the sequence appear to be bounded?

Exercise 5.9. **Definition:** The sequence $\{a_n\}$ is said to *converge to infinity*, written $\lim_{n\to\infty} a_n = \infty$, if for every real number M there is a natural number N such that if $n \geq N$ then $a_n \geq M$.

Prove that $\lim_{n\to\infty}\sqrt{n}=\infty$.

Exercise 5.10. Does the sequence $\{0, 1, 0, 2, 0, 3, 0, 4, 0, 5, \ldots\}$ converge to infinity (see Exercise 5.9 for a definition)? Why or why not?

Exercise 5.11. Suppose that $\{x_n\}$ satisfies $\lim_{n\to\infty} x_n = a > 0$. Prove that $\lim_{n\to\infty}\sqrt{x_n}=\sqrt{a}$.

Exercise 5.12. Find the limits of each of the following sequences, if they exist. Briefly justify your answer.

a) $\lim_{n\to\infty}\left(\frac{-3}{8}\right)^n$

b) $\lim_{n\to\infty}\left(\frac{-8}{3}\right)^n$

c) $\lim_{n\to\infty}\frac{3^n+4^n}{5^n}$

d) $\lim_{n\to\infty}\left(\frac{n}{2n+1}\right)^n$

e) $\lim_{n\to\infty}\frac{\sqrt[n]{2}}{\sqrt[n]{3}}$

Exercise 5.13. In each of the following, give an example of a sequence (or a pair of sequences, as needed) which satisfies the given condition or conditions, or explain why no such sequence exists.

a) a bounded sequence $\{x_n\}$ of positive terms such that $\left\{\frac{1}{x_n}\right\}$ diverges

b) divergent sequences $\{x_n\}$ and $\{y_n\}$ such that $\{x_n + y_n\}$ converges

c) sequences $\{x_n\}$ and $\{y_n\}$ where $\{x_n y_n\}$ and $\{x_n\}$ converge while $\{y_n\}$ diverges

d) bounded sequences $\{x_n\}$ and $\{y_n\}$ where $\{x_n y_n\}$ is unbounded

e) divergent sequences $\{x_n\}$ and $\{y_n\}$ such that $\{s_n y_n\}$ converges

Exercise 5.14. In each of the following give an example of a sequence (or a pair of sequences, as needed) which satisfies the given condition or conditions, or explain why no such sequence exists.

a) a sequence that contains subsequences converging to each element of the set $\{1, \frac{1}{2}, \frac{1}{3}, \frac{1}{4}, \ldots\}$

b) an unbounded sequence that has a convergent subsequence

c) an increasing sequence that diverges but has a convergent subsequence

d) a sequence that has a bounded subsequence but no subsequence that converges

e) a sequence that does not contain either 0 or 1 but has subsequences converging to each of the numbers 0 and 1

Exercise 5.15. Prove from the definition that $\{\frac{1}{n^2}\}$ is a Cauchy sequence.

Exercise 5.16. Prove that if r is a positive real number and $\lim_{n\to\infty} a_n = L$, then $\lim_{n\to\infty} ra_n = rL$.

Exercise 5.17. Prove that if $\lim_{n\to\infty} a_n = L$ and $\lim_{n\to\infty} b_n = M$ then $\lim_{n\to\infty}(a_n - b_n) = L - M$.

Exercise 5.18. Define the sequence $\{a_n\}$ by $a_n = \begin{cases} 1 + \frac{1}{n} & \text{if } 1 \le n \le 10^{10} \\ \frac{1}{n} & \text{if } 10^{10} < n \end{cases}$. Does $\{a_n\}$ converge or diverge? Give a proof of your answer.

Exercise 5.19. For each of the following series, find the sum (if it exists).

a) $\sum\limits_{k=0}^{\infty} \left(\frac{-4}{5}\right)^k$

b) $\sum\limits_{k=0}^{\infty} \frac{3^{2k}}{5^k}$

c) $\sum\limits_{k=1}^{\infty} \frac{1}{4^k}$

d) $\sum\limits_{k=1}^{\infty} \frac{1}{n^k}$, for $n \in \mathbb{N}$ with $n \ge 2$

Exercise 5.20. For each of the following series, decide whether or not it converges. Justify your conclusion. If it does converge, find the sum.

a) $1 - \frac{1}{3} + \frac{2}{9} - \frac{4}{27} + \cdots$

b) $5 + \frac{35}{5} + \frac{245}{25} + \frac{1715}{125} + \cdots$

c) $-3 - 4 - \frac{16}{3} - \frac{64}{9} - \cdots$

Exercise 5.21. Determine a closed form for the proportion of the area shaded in the diagram in Figure 5.1 of nested squares. Each side of the largest square (comprised of two white squares, a black square, and a square with subsquares) has length 1.

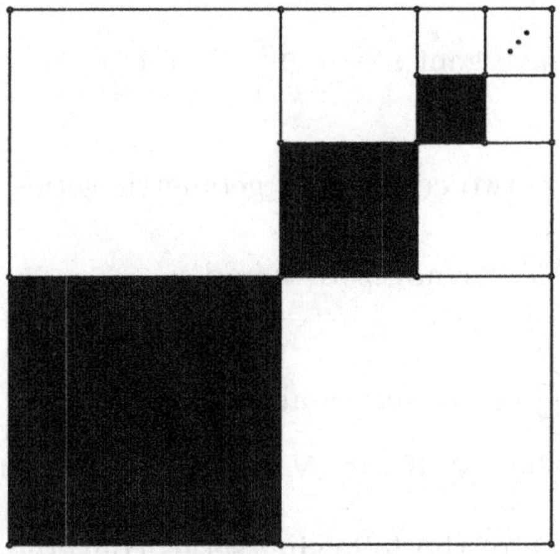

Figure 5.1

Exercise 5.22. A geometry theorem states that when connecting the midpoints of two sides of a triangle, the resulting line segment is parallel to the third side and has length half that of the third side. This fact lets us create nested similar triangles within a single triangle, as in Figure 5.2. In the figure, $\triangle ABC$ is similar to $\triangle DEF$, which is similar to $\triangle GHJ$, and so on forever beyond the practical limits of pixels. The pattern in side lengths also continues. Use this information to determine a closed form for the sum of the areas of shaded triangles in terms of the angle θ.

Figure 5.2

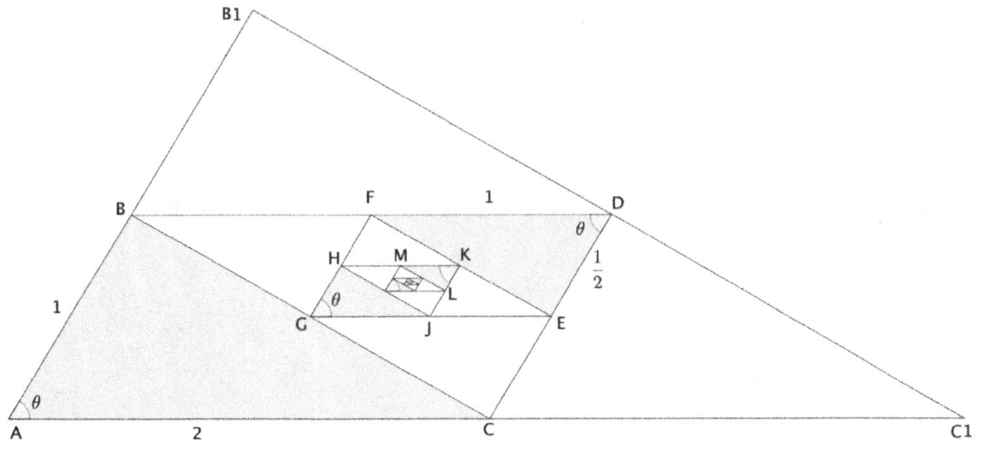

Exercise 5.23. We know that if $\sum_{k=1}^{\infty} a_k$ and $\sum_{k=1}^{\infty} b_k$ both converge then $\sum_{k=1}^{\infty}(a_k+b_k) = \sum_{k=1}^{\infty} a_k + \sum_{k=1}^{\infty} b_k$. The situation for products is not all that simple. That is, there is no simple relation between $\sum_{k=1}^{\infty} a_k$,

$\sum_{k=1}^{\infty} b_k$, and $\sum_{k=1}^{\infty} (a_k b_k)$.

a) Find an example of two divergent series $\sum_{k=1}^{\infty} a_k$ and $\sum_{k=1}^{\infty} b_k$ such that $\sum_{k=1}^{\infty} (a_k b_k)$ converges.

b) Find an example of two convergent series $\sum_{k=1}^{\infty} a_k$ and $\sum_{k=1}^{\infty} b_k$ such that $\sum_{k=1}^{\infty} (a_k b_k)$ diverges.

Exercise 5.24. Suppose we have two convergent geometric series, $\sum_{k=0}^{\infty} x^k$ and $\sum_{k=0}^{\infty} y^k$. Does it ever occur that $\sum_{k=0}^{\infty} x^k \sum_{k=0}^{\infty} y^k = \sum_{k=0}^{\infty} x^k y^k$ for non-zero x and y?

Exercise 5.25. Prove that if $\sum_{k=1}^{\infty} a_k$ converges absolutely then $\sum_{k=1}^{\infty} a_k^2$ converges absolutely. (Hint: Show that for some natural number N, if $n \geq N$ then $a_n^2 \leq |a_n|$ and use the Cauchy criterion.)

Exercise 5.26. Determine which of the following series converge and justify your answer with a convergence test.

a) $\sum_{k=1}^{\infty} \frac{3^k}{4^k+1}$

b) $\sum_{k=1}^{\infty} \frac{\ln k}{k}$

c) $\sum_{k=1}^{\infty} \frac{10^k}{k!}$

d) $\sum_{k=1}^{\infty} \frac{k^2+1}{k^4+7}$

Exercise 5.27. Suppose that the nth partial sum of $\sum_{k=1}^{\infty} a_k$ is given by $s_n = 2 + \frac{n-1}{n+1}$. Find a_1, a_2, a_{10}, and $\sum_{k=1}^{\infty} a_k$.

Exercise 5.28. For each of the following series, determine all values of x that result in a convergent series.

a) $\sum_{k=0}^{\infty} 3^k x^k$

b) $\sum_{k=0}^{\infty} \frac{k^2 x^k}{2^k}$

c) $\sum_{k=0}^{\infty} \frac{2^k x^k}{(k+1)^2}$

d) $\sum_{k=0}^{\infty} k! x^k$

Exercise 5.29. Suppose that $\sum_{k=0}^{\infty} a_k = A$. Let r be a real number. Using the partial sum definition of infinite series prove that $\sum_{k=0}^{\infty} rA$.

Exercise 5.30. Suppose that $\sum_{k=0}^{\infty} a_k = A$ and $\sum_{k=0}^{\infty} b_k = B$. Using the partial sum definition of infinite series prove that $\sum_{k=0}^{\infty} (a_k + b_k) = A + B$.

Exercise 5.31. Suppose that $\{s_n\}$ is the sequence of partial sums for the convergent infinite series $\sum_{k=1}^{\infty} a_k$. Suppose further that the infinite series $\sum_{k=1}^{\infty} b_k$ has $\{t_n\}$ as its sequence of partial sums and that for every n, $t_n = s_n + \frac{1}{n}$. Does $\sum_{k=1}^{\infty} b_k$ converge or diverge? Prove your answer.

Exercise 5.32. Let $\{a_n\} = \{a_1, a_2, a_3, \ldots\}$ be a sequence of strictly positive terms. Let $\{s_n\}$ be the partial sums for the sequence $\{a_n\}$. Prove that if there is a real number M such that $s_n \leq M$ for all n, then the infinite series $\sum_{k=1}^{\infty} a_k$ converge.

Exercise 5.33. This problem is an adaptation of #11.4.37 in Stewart's *Calculus: Early Transcendentals* (8th ed.) [8]. The decimal representation of a real number $x = 0.d_1d_2d_3d_4d_5\cdots$ where $d_i \in \{0, 1, 2, 3, \ldots, 8, 9\}$ (where we choose the terminating version, if possible) is a an infinite series.

a) Write out the infinite series equal to $x = 0.d_1d_2d_3d_4d_5\cdots$.

b) Find a converging infinite series that is related to the one equal to x, but slightly larger.

c) Prove that this second series converges.

d) Apply one of the Comparison Tests (Theorems 5.32 and 5.41) to conclude that that the series for x converges.

As a consequence of your work, the decimal representation of a number in $[0, 1]$ converges and by extension decimal representation as a system for all real numbers is a consistent, well-defined system.

Exercise 5.34. Suppose that (x_n) is a decreasing sequence that converges to 0. Prove, using the definition, that the sequence (y_n) where $y_n = x_1 - x_2 + x_3 - x_4 + \cdots + (-1)^{n+1}x_n$ is a Cauchy sequence. Note that this is an alternate approach to proving the Alternating Series Test (Theorem 5.35).

Exercise 5.35. In this problem, you will work through a proof of the Limit Comparison Test (Theorem 5.41).

Step 1. Suppose that $\{a_n\}$ and $\{b_n\}$ are sequences of positive real numbers and that $\lim_{n\to\infty} \frac{a_n}{b_n} = L$ with $0 < L < \infty$. Show that there is a natural number N such that if $n \geq N$ then $a_n \leq 2Lb_n$.

Step 2. Now use the Cauchy Convergence Criterion (Theorem 5.31) to show that if $\sum_{k=1}^{\infty} b_k$ converges then $\sum_{k=1}^{\infty} a_k$ must converge.

Step 3. Show that there is a natural number M such that if $n \geq M$ then $\frac{Lb_n}{2} \leq a_n$.

Step 4. Modify Step 2 to show that if $\sum_{k=1}^{\infty} a_k$ converges then $\sum_{k=1}^{\infty} b_k$ must converge.

Exercise 5.36. In this problem, you will be stepped through a proof of Case 1) of the Ratio Test (Theorem 5.42). Suppose that $\lim_{n\to\infty} \left|\frac{a_{n+1}}{a_n}\right| = \lambda < 1$ for the infinite series $\sum_{k=1}^{\infty} a_k$. Follow the steps outlined below to prove the conclusion, that $\sum_{k=1}^{\infty} a_k$ converges absolutely.

Step 1. Show that for there is a natural number N and a real number μ satisfying $\lambda < \mu < 1$ such that if $n \geq N$ then $\left|\frac{a_{n+1}}{a_n}\right| \leq \mu$.

Step 2. Show htat the series $\sum_{k=1}^{\infty} \mu^k$ converges absolutely.

Step 3. Show that $|a_{N+k}| \leq |a_N|\mu^k$.

Step 4. Use the comparison test on the series $\sum_{k=1}^{\infty} |a_k|$ and $\sum_{k=1}^{\infty} \mu_k$ to complete the proof.

Chapter 6

The Topology of the Real Numbers

Topology is the study of space at the most abstract level. When applied to the real numbers it gives us many important tools to understand the real numbers and functions of them. Topology is based on the basic notions of "open set" and "limit point". Our task in this chapter is to develop these notions in order to define the limit of a function. We begin with the definition of an epsilon neighborhood of a point.

Definition 6.1 (Epsilon Neighborhood). Let x and $\varepsilon > 0$ be real numbers. The *ε-neighborhood of x* is the set $\mathrm{N}_\varepsilon(x)$ defined by:

$$\mathrm{N}_\varepsilon(x) = \{y \in \mathbb{R} \mid |x - y| < \varepsilon\} = (x - \varepsilon, x + \varepsilon).$$

Example 6.2.

1. The 1-neighborhood of 0 is the set of all real numbers less than one unit away from 0. Namely: $\mathrm{N}_1(0) = (-1, 1)$.
2. The 3-neighborhood of 5 is $\mathrm{N}_3(5) = (2, 8)$.
3. The 0.01-neighborhood of 2 is $\mathrm{N}_{0.01}(2) = (1.99, 2.01)$.

Definition 6.3 (Open Set). A set U of real numbers is said to be an *open set* if for each $x \in U$ there is an $\varepsilon > 0$ such that $\mathrm{N}_\varepsilon(x) \subset U$. Note that this condition has to hold for every point x in the set U but that for each x only one ε is required.

Example 6.4.

1. $(0, 1)$ is an open set.

   To prove this let x be an arbitrary point in $(0, 1)$. We need to find a small open interval centered on x that is entirely contained in $(0, 1)$. Let $\varepsilon = \min\{x, 1 - x\}$. Thus ε is the distance from x to the closest endpoint 0 or 1. We need to show that $\mathrm{N}_\varepsilon(x) \subset U$ for every x.

   Case 1. Suppose that $0 < x \leq \frac{1}{2}$. Then $\varepsilon = x$. Since $2x \leq 1$,

   $$\mathrm{N}_\varepsilon(x) = (x - \varepsilon, x + \varepsilon) = (0, 2x) \subset (0, 1).$$

   Case 2. Suppose $\frac{1}{2} < x < 1$. Then $\varepsilon = 1 - x$. Since $2x - 1 \geq 0$,

   $$\mathrm{N}_\varepsilon(x) = (x - (1 - x), x + (1 - x)) = (2x - 1, 1) \subset (0, 1).$$

2. The empty set, $\emptyset$, is an open set.

   The condition is satisfied for every x in the empty set, of which there are none.

3. The set $\mathbb{R}$ is an open set.

   For each $x \in \mathbb{R}$, let $\varepsilon = 1$. Then $\mathrm{N}_\varepsilon(x) = (x-1, x+1) \subset \mathbb{R}$.

The fact that $\mathbb{R}$ and $\emptyset$ are open sets is surprisingly important.

Definition 6.5 (Interior Point of a Set). An element $x \in U$ is an *interior point of the set* U if there is an $\varepsilon > 0$ satisfying $\mathrm{N}_\varepsilon(x) \subset U$. Every point in an open set is an interior point. The *interior of a set* U is the set of all interior points of U. We denote this by $U^\circ = \{x \in U \mid \exists \varepsilon > 0 \text{ such that } \mathrm{N}_\varepsilon(x) \subset U\}$.

Theorem 6.6. A set U is open if and only if $U = U^\circ$.

For the proof, simply read the two relevant definitions.

Topology is based on the following two properties of open sets, properties that are axioms for a general topological space.

Theorem 6.7.

(1) Let $\{U_\alpha \mid \alpha \in A\}$ be any collection of open sets, of any cardinality. Then $\bigcup_{\alpha \in A} U_\alpha$ is an open set.

(2) Let $\{U_1, U_2, \ldots, U_n\}$ be any finite collection of open sets. Then $\bigcap_{k=1}^n U_k = U_1 \cap U_2 \cap \cdots \cap U_n$ is an open set.

Proof. For part (1): Suppose that $x \in \bigcup_{\alpha \in A} U_\alpha$. Then by the definition of union $x \in U_\beta$ for a particular $\beta \in A$. Thus there is an $\varepsilon > 0$ such that $\mathrm{N}_\varepsilon(x) \subset U_\beta \subset \bigcup_{\alpha \in A} U_\alpha$.

For part (2): This case is somewhat more difficult than 6.7.(1). Let $x \in \bigcap_{k=1}^n U_k$. Since for each k, $1 \le k \le n$, the set U_k is open, there is an $\varepsilon_k > 0$ such that $\mathrm{N}_{\varepsilon_k}(x) \subset U_k$. Since there are finitely many of these ε_k, we can find a smallest one. Let $\varepsilon = \min\{\varepsilon_1, \varepsilon_2, \ldots, \varepsilon_k\} > 0$. Since $\varepsilon \le \varepsilon_k$ for each k, we have that $\mathrm{N}_\varepsilon(x) \subset \mathrm{N}_{\varepsilon_k}(x) \subset U_k$. Thus the set $\mathrm{N}_\varepsilon(x)$ is contained in the intersection of the sets, that is $\mathrm{N}_\varepsilon(x) \subset \bigcap_{k=1}^n U_k$. Thus the intersection is an open set. □

Example 6.8.

1. Consider the following collection of open intervals:

$$\{U_n\} = \left\{\left(-1 + \frac{1}{n}, 1 - \frac{1}{n}\right)\right\} = \left\{\emptyset, \left(\frac{-1}{2}, \frac{1}{2}\right), \left(\frac{-2}{3}, \frac{2}{3}\right), \ldots\right\}.$$

   The union of all these sets is $(-1, 1)$, an open set as promised by the theorem. Their intersection (the empty set) is also open but that is not promised by the theorem.

2. Consider the following different collection of open intervals:

$$\{U_n\} = \left\{\left(0, 1 + \frac{1}{n}\right)\right\} = \left\{(0, 2), \left(0, \frac{3}{2}\right), \left(0, \frac{4}{3}\right), \ldots\right\}.$$

Taking the intersection of this infinite collection of open sets yields:

$$\bigcap_{k=1}^{\infty}\left(0, 1+\frac{1}{k}\right) = (0, 1],$$

which is not an open set.

The usefulness of open sets will become apparent when we have defined the complementary concept of closed set. We defined open set and then interior point. For closed sets we define limit points first (the complementary notion to interior point) and then define closed set.

Definition 6.9 (Limit Point of a Set). Let K be a set of real numbers. The number x is a *limit point of* K if for every $\varepsilon > 0$ the intersection $K \cap \mathrm{N}_\varepsilon(x)$ contains a point different from x.

This definition is rather subtle. There are many things it does not say. For example it does not say that x is in the set K. It also does not say that x is *not* in K. Also notice the "for every $\varepsilon > 0$," which differs from the "there exists $\varepsilon > 0$" in the definition of open set.

A few examples will bring out the difficulties.

Example 6.10 (Limit Points).

1. The number 1 is a limit point of $(0, 1)$.

   Let $\varepsilon > 0$ be given. We need to find an element (different from 1) that is in $(0,1) \cap \mathrm{N}_\varepsilon(1) = (0,1) \cap (1-\varepsilon, 1+\varepsilon)$. If $\varepsilon \geq 1$, we can choose 0.5 in this intersection (as is any other point in $(0,1)$). If $\varepsilon < 1$ then $(0,1) \cap (1-\varepsilon, 1+\varepsilon) = (1-\varepsilon, 1)$. Then $1 - \frac{\varepsilon}{2}$ is in the intersection and hence we have proved that 1 is a limit point of $(0, 1)$.

2. The number 1 is a limit point of $[0, 1]$. The proof is the same as the above example.

3. The number 0 is a limit point of the set $S = \left\{\frac{1}{n} \middle| n \in \mathbb{N}\right\} = \left\{1, \frac{1}{2}, \frac{1}{3}, \ldots\right\}$.

   Let $\varepsilon > 0$ be given. By the Archimedean Principle (Theorem 3.9), there is a natural number n such that $0 < \frac{1}{n} < \varepsilon$. Notice that $\frac{1}{n}$ is in $\mathrm{N}_\varepsilon(0) \cap S = (-\varepsilon, \varepsilon) \cap S = \left\{\frac{1}{m} \middle| \mid m \in \mathbb{N} \text{ and } \frac{1}{m} < \varepsilon\right\}$.

4. The number $\sqrt{2}$ is a limit point of $\mathbb{Q}$, the rational numbers.

   Again let $\varepsilon > 0$ be given. By the Density Theorem (Theorem 3.10) there is a rational number between the two real numbers $\sqrt{2} - \varepsilon$ and $\sqrt{2} + \varepsilon$. That rational number is all we need for the proof.

Definition 6.11 (Closed Set). A set K is *closed* if it contains all its limit points. That is, the set K is closed if whenever x is a limit point of the set K then $x \in K$.

Again examples help one to understand this definition.

Example 6.12 (Closed Sets).

1. The number 1 is a limit point of $(0, 1)$ and of $[0, 1]$. Clearly $(0, 1)$ is not a closed set since it does not contain one of its limit points, namely 1. The interval $[0, 1]$ is a closed set but that takes a bit of proof. We know that it contains one of its limit point (namely, 1) but what about all the others?

2. The empty set is a closed set. It doesnt have any limit points so it must contain them all.

3. The set of real numbers is a closed set since it contains all possible limit points.

4. The rational numbers is not a closed set. We saw that $\sqrt{2}$ is a limit point of $\mathbb{Q}$ but it is not in $\mathbb{Q}$.

5. Let $S = \{1, \frac{1}{2}, \frac{1}{3}, \dots\}$. The number 0 is not in S but we showed that it is a limit point of S, thus S is not closed. The set $S \cup \{0\}$ is a closed set.

Open and closed sets are closely related. In fact the complement of an open set is a closed set and vice versa. To show this we need lemma from the theory of sets.

Lemma 6.13. Let A and B be sets. Then $A \subset B$ if and only if $A \cap B^c = \emptyset$.

Proof. Suppose that $A \subset B$. Then every element of A is an element of B and hence cannot be in B^c. Thus $A \cap B^c = \emptyset$. The other direction is just as easy. □

Theorem 6.14. The complement of an open set is a closed set and the complement of a closed set is an open set.

Proof. Let U be an open set. Let $x \in U$. We need to show that x cannot be a limit point of U^c. Since U is open there exists an $\varepsilon > 0$ such that $\mathrm{N}_\varepsilon(x) \subset U$. By the preceding lemma $\mathrm{N}_\varepsilon(x) \cap U^c = \emptyset$. Thus x is not a limit point of U^c. Hence all the limit points of U^c are already in U^c and thus it is a closed set.

Now let K be a closed set. Let $x \in K^c$. By the definition of closed set x cannot be a limit point of K. So $\mathrm{N}_\varepsilon(x) \cap K = \emptyset$ for some $\varepsilon > 0$. Again by the lemma $\mathrm{N}_\varepsilon(x) \subset K^c$ and hence K^c is an open set. □

Theorem 6.15.

(1) Let $\{K_\alpha \mid \alpha \in A\}$ be any collection of closed sets. Then $\bigcap_{\alpha \in A} K_\alpha$ is a closed set.

(2) Let $\{K_1, K_2, \dots, K_n\}$ be a finite collection of closed sets. Then $\bigcup_{k=1}^n K_k = K_1 \cup K_2 \cup \dots \cup K_n$ is a closed set.

Proof. Notice that for closed sets the roles of union and intersection are reversed from their roles with open sets (Theorem 6.7). The proof of this theorem is simply the application of the DeMorgan Laws (Theorem 2.13) to Theorem 6.7. We will prove (2).

The set $\{K_1^c, K_2^c, \dots, K_n^c\}$ is a finite collection of open sets and hence its intersection is open, that is $\bigcap_{k=1}^n K_k^c = K_1^c \cap \dots \cap K_n^c$ is an open set. Thus the complement of the intersection is a closed set. Which closed set is it?

$$\left(\bigcap_{k=1}^{n} K_k^c\right)^c = \bigcup_{k=1}^{n} (K_k^c)^c = \bigcup_{k=1}^{n} K_k$$

This finishes the proof. □

Open sets are modeled on open intervals. One can prove that every open set is the union of countably many open intervals. Closed sets are much harder to characterize. Here is an example of how odd closed sets can be.

Example 6.16. Let $\varepsilon > 0$ be a very small real number. Let $\{x_1, x_2, x_3, \dots\}$ be an enumeration of the rational numbers. Thus every rational number is in the list in exactly one position in the sequence. For each natural number n let I_n be the open interval $\left(x - \frac{\varepsilon}{2^{n+2}}, x + \frac{\varepsilon}{2^{n+2}}\right)$. Notice that the length of this interval is $\frac{\varepsilon}{2^{n+1}}$. Consider the open set $I = \bigcup_{k=1}^{\infty} I_k$. The set I contains every rational number since each rational is the center point of an interval contained in I. The sum of the lengths of all the intervals is less than or equal to

$$\sum_{k=1}^{\infty} \frac{\varepsilon}{2^{k+1}} = \varepsilon \sum_{k=1}^{\infty} \left(\frac{1}{2}\right)^{k+1} = \varepsilon \sum_{k=2}^{\infty} \left(\frac{1}{2}\right)^{k} = \frac{\varepsilon}{2}.$$

Thus I is a very small open set, of total length at most $\frac{\varepsilon}{2}$. Its complement is a closed set whose total length is infinite, a huge set of real numbers. It contains no rational numbers and is a closed set. Since it contains no rational numbers it contains no closed intervals of positive length. Every closed interval of positive length has to contain a rational number by the Density Theorem (Theorem 3.10).

The notion of limit point and convergent sequence come together in the following theorem which gives a criterion for a point being a limit point of a set.

Theorem 6.17. The element x is a limit point of set K if and only if there exists a sequence $\{x_k\}$ contained in set K satisfying two conditions:

(1) $x_k \neq x$ for all $k \in \mathbb{N}$, and

(2) $\lim_{n\to\infty} x_n = x$.

Proof. Assume that x is a limit point of K. For each natural number n there is a point x_n of K in the $\frac{1}{n}$-neighborhood of x that is different from x. That is $x_n \in \mathrm{N}_{\frac{1}{n}}(x) \cap K$ and $x_n \neq x$. These x_n form a sequence $\{x_n\}$. All that is left to show is that $\lim_{n\to\infty} x_n = x$.

Let $\varepsilon > 0$ be given. By the Archimedean Principle (Theorem 3.9) there is a natural number N such that $\frac{1}{N} < \varepsilon$. If $n \geq N$, then $\frac{1}{n} \leq \frac{1}{N} < \varepsilon$ and $x_n \in \mathrm{N}_{\frac{1}{n}}(x) \subset \mathrm{N}_{\varepsilon}(x)$. Thus $|x_n - x| < \varepsilon$ for all $n \geq N$ and we have convergence.

Now assume the existence of a sequence $\{x_n\}$ with the specified properties and show that x is a limit point of K. We need to show that for each $\varepsilon > 0$ there is a point of K different from x in the neighborhood $\mathrm{N}_{\varepsilon}(x)$. Since $\lim_{n\to\infty} x_n = x$ there are always terms in the sequence within ε of x. These terms are never equal to x, thus x is a limit point of K. □

Example 6.18. Show that 1 is a limit point of $(0,1)$ using the above theorem.

Let $x_n = 1 - \frac{1}{n+1}$. Then each term of the sequence $\{x_n\} = \left\{1 - \frac{1}{n+1}\right\}$ is in $(0,1)$, no term equals 1, and $\lim_{n\to\infty} 1 - \frac{1}{n+1} = \lim_{n\to\infty} \frac{n}{n+1} = 1$.

Definition 6.19 (The Closure of a Set)**.** Let A be a set of real numbers. The *closure of* A, denoted by $\overline{A}$, is the set consisting of A and its limit points. That is, if L is the set of limit points of A then $\overline{A} = A \cup L$.

Example 6.20.

1. $\overline{(0,1)} = [0,1]$

2. $\overline{\mathbb{Q}} = \mathbb{R}$

3. $\overline{\{1,2,3\}} = \{1,2,3\}$

Theorem 6.21. For any set A, the closure $\overline{A}$ is a closed set. Furthermore A is closed if and only if $A = \overline{A}$.

Proof. All we need to show is that when we create $\overline{A}$, we are not creating any new limit points. Let x be a limit point of $\overline{A}$. We need to show that $x \in \overline{A}$. Let $\varepsilon > 0$ be given. Since x is a limit point of $\overline{A}$, there is a point $y \in \overline{A}$, different from x, that is within ε of x. We need to show that we can always choose y so that it is in A. Then x will be a limit point of A and hence in the closure of A. Suppose our first choice of y is a limit point of A but not an element of A. Let $\delta = |x - y| < \varepsilon$ and let $\omega = \varepsilon - \delta > 0$. Since y is a limit point of A, there is a point z in A, different from y, such that $|y - z| < \omega$, in other words $z \in \mathrm{N}_\omega(y)$ and $z \neq y$. By the Triangle Inequality (Theorem 2.11),

$$|z - x| \leq |z - y| + |y - x| < \omega + \delta = \varepsilon.$$

Thus z is in the ε-neighborhood of x and z is in A. Thus x is a limit point of A.

The second part of the theorem is quite easy to prove. □

We now have all the basic facts about open and closed sets. There is a common confusion about these concepts. If a set is not open it is not necessarily closed. There are many (infinitely many) sets that are neither open nor closed. Table 6.1 gives some examples of the relation between open and closed.

	Closed	Not Closed
Open	$\emptyset$, $\mathbb{R}$	$(0,1)$
Not Open	$[0,1]$	$(0,1]$, $[0,1)$

Table 6.1

Thus there are sets that are neither open nor closed and sets (exactly two of them) that are both open and closed. Open and closed are complementary concepts, not opposite concepts.

Definition 6.22 (Isolated Point). A point x in set A is an *isolated point of* A if it is not a limit point of A.

Every point of A is either a limit point or an isolated point. For example the set $A = \{1,2,3\}$ consists of only isolated points.

We have one more topological (space) concept to define, that of compactness.

Definition 6.23 (Compact Set). A set K is *compact* if every sequence, $\{x_n\}$, contained in K has a subsequence that converges to a point in K. (In some texts this is called "sequentially compact" and "compact" has a different definition. For the real numbers the two define equivalent concepts.)

Example 6.24. The interval $[0,1]$ is compact.

Proof. Let $\{x_n\}$ be a sequence in $[0,1]$. Since $0 \leq x_n \leq 1$ for each n, the sequence is bounded. Thus by the Bolzano-Weierstrass Theorem (Theorem 5.21) it contains a convergent subsequence. Call the subsequence $\{y_n\}$ and suppose that its limit is y. If for some n, $y_n = y$, then y is in $[0,1]$ and hence $[0,1]$ is compact. So suppose that $y \neq y_n$. Then by the earlier Theorem 6.17, y is a limit point of $[0,1]$ and since that set is closed, y is in $[0,1]$. We have shown compactness. □

Showing that a set is compact directly is often very difficult. For the real numbers, however, there is an easy characterization of compact sets.

Theorem 6.25 (Heine-Borel Theorem). A set is compact if and only if it is closed and bounded.

Proof. Assume that K is a compact set (according to the definition). We need to show that it is both closed and bounded. To prove that K is closed is easier so we start with that. Let x be a limit point of K. We need to show that $x \in K$. Since it is a limit point we can use Theorem 6.17 to find a sequence $\{x_n\} \subset K$ satisfying $x \neq x_n$ and $\lim_{n\to\infty} x_n = x$. Since by the compactness of K this sequence has a subsequence that converges to a point of K and we know that every subsequence of a convergent sequence converges to the limit of the sequence, we know that x is in K.

To prove that K is bounded will be accomplished by contrapositive. We assume that K is not bounded above. (The not bounded below case is practically the same). Let x_1 be an element of K. Since K is not bounded above, x_1 cannot be an upper bound of K. Thus there must be $x_2 \in K$ satisfying $x_1 < x_2$. Further we can choose x_2 to be greater than 2. Continuing in this fashion we can choose $x_n \in K$ satisfying $x_{n-1} < x_n$ and $n < x_n$ for each natural number n. We have thus constructed an unbounded sequence $\{x_n\}$ that has no bounded subsequences. Thus it has no convergent subsequences. Hence K is not compact.

Now for the opposite implication, we assume that K is both closed and bounded. This part of the proof mimics the flow of the above example 6.24.

Let $\{x_n\}$ be a sequence contained in K. Since K is bounded so is the sequence $\{x_n\}$. By the Bolzano-Weierstrass Theorem (Theorem 5.21) the sequence has a convergent subsequence, call it $\{y_n\}$, which converges to a point y. We need to show that $y \in K$. This holds because K is closed. If y occurs in the subsequence then it is already in K and we are done. Let us assume that $y_n \neq y$ for all n. Then by Theorem 6.17, y is a limit point of K. Since K is closed and hence contains all its limit points it follows that $y \in K$. □

In what follows we will define what is called the Cantor Set, an interesting closed set of real numbers. It was discovered by Georg Cantor while he was studying the discontinuities of functions defined by Fourier Series. We begin by defining a sequence of closed sets.

Let $C_0 = [0,1]$. From C_0 we remove its middle third to yield a set called C_1. That is,

$$C_1 = C_0 - \left(\frac{1}{3}, \frac{2}{3}\right) = [0,1] - \left(\frac{1}{3}, \frac{2}{3}\right) = \left[0, \frac{1}{3}\right] \cup \left[\frac{2}{3}, 1\right].$$

This is the union of two closed sets and hence is closed. We then repeat the process by removing the middle thirds of the intervals remaining. The middle thirds of $[0, \frac{1}{3}]$ and $[\frac{2}{3}, 1]$ are $(\frac{1}{9}, \frac{2}{9})$ and $(\frac{7}{9}, \frac{8}{9})$, respectively. Note that we are removing open intervals in each case. Thus we have that

$$C_2 = C_1 - \left(\left(\frac{1}{9}, \frac{2}{9}\right) \cup \left(\frac{7}{9}, \frac{8}{9}\right)\right) = \left[0, \frac{1}{9}\right] \cup \left[\frac{2}{9}, \frac{1}{3}\right] \cup \left[\frac{2}{3}, \frac{7}{9}\right] \cup \left[\frac{8}{9}, 1\right].$$

With each step the closed sets C_n get more and more complicated. It is easy to see that C_n consists of 2^n disjoint closed intervals each of width $\left(\frac{1}{3}\right)^n$. For example,

$$C_3 = \left[1, \frac{1}{27}\right] \cup \left[\frac{2}{27}, \frac{1}{9}\right] \cup \left[\frac{2}{9}, \frac{7}{27}\right] \cup \left[\frac{8}{27}, \frac{1}{3}\right] \cup \left[\frac{2}{3}, \frac{19}{27}\right] \cup \left[\frac{20}{27}, \frac{7}{9}\right] \cup \left[\frac{8}{9}, \frac{25}{27}\right] \cup \left[\frac{26}{27}, 1\right].$$

To form the Cantor Set, C, we take the intersection of all these sets:

$$C = C_0 \cap C_1 \cap C_2 \cap \cdots = \bigcap_{n=0}^{\infty} C_n.$$

This set has several names — the *Cantor Ternary Set* (since it involves the removal of middle thirds) and *Cantor Dust* (since in some sense we will not define, there are very few points in C) are common alternatives.

The set C is a very difficult set to try to picture in your mind. We can say that as an intersection of closed sets it is closed and, further, since each of the endpoints of each of the closed intervals stays in the set, C is non-empty. For example C contains $0, 1, \frac{1}{3}, \frac{1}{9}, \frac{1}{27}, \ldots$ and much more. Thus we can see that C is at least a countably infinite set. We will show that C is actually uncountable.

All the points listed so far are of the form $\frac{k}{3^n}$ but we shall show that $\frac{1}{4}$ is in C. To explore C we are going to recall base 3 expansions of numbers. For example,

$$\frac{1}{3} = 0.1_3 \quad \text{and} \quad \frac{5}{27} = 0.012_3.$$

Also,

$$0.012_3 = 0 + \frac{0}{3} + \frac{1}{3^2} + \frac{2}{3^3} = \frac{1}{9} + \frac{2}{27} = \frac{5}{27}.$$

Every real number in $[0,1]$ has a base three expansion and some have two different expansions. Recall that in base 10, we observed $1.0 = 0.99999\cdots$. The same phenomenon occurs in base three. For example $\frac{1}{3} = 0.1_3 = 0.02222\cdots_3$. This occurs for every real number that has a terminating base three expansion, namely those rational numbers whose denominators (in lowest terms) are powers of 3. For those numbers we will always choose to use the repeating, non-terminating expansion. For example $1.0 = 0.22222\cdots_3$. It is helpful at this point to recall the Geometric Series.

$$1 = 0.2222\cdots_3 = \frac{2}{3} + \frac{2}{9} + \frac{2}{27} + \cdots = \sum_{k=1}^{\infty} \frac{2}{3^k} = 2\sum_{k=1}^{\infty} \frac{1}{3^k} = 2\left(\frac{\frac{1}{3}}{1-\frac{1}{3}}\right) = 2\left(\frac{1}{2}\right) = 1$$

Now consider the Cantor Set C and the sets we remove. At our first step we remove the open interval $\left(\frac{1}{3}, \frac{2}{3}\right)$. Every number in this set has a base three expansion whose first digit after the point is 1, since they are all greater than $\frac{1}{3}$ and less than $\frac{2}{3}$. The number $\frac{1}{3}$ itself has a base three expansion of 0.1_3 but we have chosen to use the expansion $0.02222\cdots_3$. Thus removing the first open interval is equivalent to removing all numbers that have a 1 in the first base 3 position. It is not hard to see that when we remove the second pair of open intervals we are removing all the numbers that have a 1 in their second base three position. (Again, we are using our convention of picking the repeating form rather than the terminating form in each case.)

Thus C can be viewed as the set of all real numbers in $[0,1]$ whose base three expansions consist of only 0s and 2s. All the numbers that have a 1 in their expansion have been removed.

For example the number $0.02020202\cdots_3$ is in C. What is that number?

$$0.02020202\cdots_3 = \frac{2}{9} + \frac{2}{81} + \frac{2}{729} + \cdots = 2\sum_{k=1}^{\infty}\left(\frac{1}{9}\right)^k = 2\left(\frac{\frac{1}{9}}{1-\frac{1}{9}}\right) = 2\left(\frac{1}{8}\right) = \frac{1}{4}$$

Thus we have shown that $\frac{1}{4} \in C$.

Theorem 6.26. The Cantor Set C is uncountable.

Proof. The set C consists of all real numbers in $[0, 1]$ that have base three expansions consisting of just the digits 0 and 2. The numbers in $[0, 1]$ also have base 2 expansions all of whose digits are 0 or 1. (We choose $1.0 = 0.1111\cdots_2$). Consider the function $f : [0, 1] \to C$ defined by $f(x)$ equals the base three expansion formed by doubling all the digits of the base 2 expansion of the numbers in $[0, 1]$. For example $f(0.11001100\cdots_2) = 0.22002200\cdots_3$. It is easy to show that f is one to one and onto and hence that C has the same cardinality as the interval $[0, 1]$, hence C is uncountable. $\square$

6.1 Exercises

Exercise 6.1. Suppose that A and B are non-empty sets of real numbers and that x is a limit point of $A \cup B$. Prove that x is a limit point of A or of B.

Exercise 6.2. Prove: A is closed if and only if $A = \overline{A}$.

Exercise 6.3. Prove that if a set is closed and bounded then it is compact.

Exercise 6.4. Prove that the interval $(2, 4)$ is an open set.

Exercise 6.5. Prove that the interval $[2, 4]$ is a closed set.

Exercise 6.6. Prove that $(7, \infty)$ is an open set.

Exercise 6.7. Let $A = \left\{1 + \frac{(-1)^n}{n} \middle| n \in \mathbb{N}\right\}$. Find all the limit points of A. Is A closed? Find $\overline{(A^c)}$ and $(\overline{A})^c$.

Exercise 6.8. Find a set with exactly three limit points.

Exercise 6.9. Find an irrational number that is an interior point of the set of rational numbers.

Exercise 6.10. Prove that the empty set is compact.

Exercise 6.11. Find an example of a compact set which does not contain its least upper bound.

Exercise 6.12. Prove that if A and B are compact then so are $A \cap B$ and $A \cup B$.

Exercise 6.13. Find a collection of compact sets such that the intersection of all the sets is not compact.

Exercise 6.14. Which of the following sets are compact?

a) $\{x | -1 < x^2 \leq 4\}$

b) $\mathbb{Q} \cap [0, 1]$

c) $\left\{1, \frac{1}{2}, \frac{1}{3}, \frac{1}{4}, \ldots\right\}$

d) $\left\{1, \frac{1}{2}, \frac{2}{3}, \frac{3}{4}, \ldots\right\}$

e) $[-1, 0) \cup (0, 1]$

Exercise 6.15. Prove that every finite set is compact.

Exercise 6.16. Find a nested set of non-empty closed sets $K_1 \supset K_2 \supset K_3 \supset \cdots$ satisfying $\bigcap_{i=1}^{\infty} K_i = \emptyset$ (if possible).

Exercise 6.17. Let $A = \left\{\frac{m}{2^m} \middle| m, n \in \mathbb{Z}\right\}$. The number $\frac{1}{3}$ is a limit point of A. Find a sequence in A that converges to $\frac{1}{3}$. (Hint: Find an x such that $\sum_{i=1}^{\infty} x^n = \frac{1}{3}$. This gives a sequence which converges to $\frac{1}{3}$.)

Exercise 6.18. Find all the values of $n = 0, 1, 2, \ldots, 13$ such that $\frac{n}{13}$ is an element of the Cantor set, $\mathcal{C}$.

Exercise 6.19. Prove that the following sets are not open sets.

a) The rational numbers, $\mathbb{Q}$

b) $\{1, \frac{1}{2}, \frac{1}{3}, \dots\}$

c) $(2, 7]$

Exercise 6.20. Suppose $\varepsilon > \delta > 0$. Let $A = \mathrm{N}_\varepsilon(x)$ for some given real number x. Find $A \cup B$ and $A \cap B$. Then find ε and x if $\mathrm{N}_\varepsilon(x) = \mathrm{N}_3(4) \cup \mathrm{N}_5(3)$.

Exercise 6.21. Let $\{x_1, x_2, x_3, \dots\}$ be an enumeration of the rational numbers. For each natural number n let $\varepsilon_n = \frac{1}{2^n}$ and let $U_n = \mathrm{N}_{\varepsilon_n}(x_n)$. Finally, let $U = \bigcup_{n=1}^{\infty} U_n$, which is an open set. What is the sum of the lengths of all the sets U_n? Are there any rational numbers in U^c? Why or why not?

Exercise 6.22. Let U be an open set with $x \in U$. Suppose that $\{x_n\}$ is a sequence that converges to x. Prove that at most finitely many terms of the sequence are not in U.

Exercise 6.23. Let x be a real number that is not in the set A. Suppose there is a sequence $\{x_n\} \subset A$ such that $\lim_{n\to\infty} x_n = x$. Prove that x is a limit point of set A.

Chapter 7

Limits and Continuity

We defined our first limit, the limit of a sequence, earlier. We now move to a more complex and subtle limit, the limit of a function. Since a sequence was defined as a function it may seem that we are doing nothing new, however, sequences are functions with a domain of the natural numbers. We wish to deal with functions on general subsets of the real numbers. The topology presented in the last chapter will enter the definition. This is the limit used in calculus to define the derivative. However, our definition will differ from that given in calculus in an important way.

Definition 7.1 (Limit of a Function). Let $f : A \to \mathbb{R}$ be a function and let c be a limit point of the set A. We say that the *limit of f as x approaches c is L*, written $\lim_{x\to c} f(x) = L$, if for every $\varepsilon > 0$, there is a $\delta > 0$ such that if $0 < |x - c| < \delta$ and $x \in A$ then $|f(x) - L| < \varepsilon$.

Remark 7.2. There are several points about this definition that need to be brought up before we try to compute a limit.

1. The value of δ depends on ε, that is given ε it is our job to find a δ that satisfies the definition. Since we want to do this for all ε at once, what we really want is to express δ as some function, say $\delta = a(\varepsilon)$.

2. The c in the definition is a limit point of the domain of f but not necessarily in the domain of f. Thus $f(c)$ may or may not exist, and if it does exist its value may be different from the limit value L.

3. In calculus A is usually an open interval containing c or with c left out. Thus the limit from calculus is a two sided limit. In our definition A can be a much more general set. For example if f is a function on $A = \left\{1, \frac{1}{2}, \frac{1}{3}, \frac{1}{4}, \dots\right\}$ then the limit $\lim_{x\to 0} f(x)$ may exist, since 0 is a limit point of A. But since A has no other limit points, there are no other limits involving f.

Example 7.3.

1. Let $f : \mathbb{R} \to \mathbb{R}$ be defined by $f(x) = x$ and let c be any real number. Since c is a limit point of the domain $\mathbb{R}$, the limit $\lim_{x\to c} f(x) = \lim_{x\to c} x$ may be defined. In fact it is. We will show that $\lim_{x\to c} x = c$. This is not surprising since on an intuitive level it says that the limit of x as x approaches c is c.

   Let $\varepsilon > 0$ be given. Let $\delta = \varepsilon$ be the functional relation between these two quantities. If x is a real number then it is in the domain of our function. Now suppose that $0 < |x - c| < \delta$. Neglecting the left inequality and using the fact that $\delta = \varepsilon$ we have that $|f(x) - c| = |x - c| < \delta = \varepsilon$ or just $|x - c| < \varepsilon$ and we are done.

2. Example 7.3.1. was very easy due to the simplicity of f. Making f more complex we show that $\lim_{x\to 2} x^2 = 4$. Again we assume that the domain of $f(x) = x^2$ is the set of all real numbers and thus 2 is a limit point of the domain. Let $\varepsilon > 0$ be given. We need to find $\delta > 0$ such that if $0 < |x-2| < \delta$ then $|x^2 - 4| < \varepsilon$. Factoring $x^2 - 4$ into $(x-2)(x+2)$, we find that $x+2$ defines the relation between ε and δ. Since x varies, we dont really know how big $|x+2|$ is. We control how big it can get by fixing a possible value of δ, say $\delta = 1$. Then $|x-2| < \delta = 1$ yields $-1 < x-2 < 1$ or $1 < x < 3$. We really want to control the size of $|x+2|$. With our condition that $1 < x < 3$ we have $3 < |x+2| < 5$. The 5 is the number of interest. Here is the proof that $\lim_{x\to 2} x^2 = 4$.

   Given $\varepsilon > 0$, let $\delta = \min\left\{1, \frac{\varepsilon}{5}\right\}$. If $0 < |x-2| < \delta$ then we know that $|x+2| < 5$ by our calculation above since we are assuming that $\delta \leq 1$. Then $|x^2 - 4| = |x-2||x+2| < \delta \cdot 5 \leq \varepsilon$. And we are done.

   What follows is not part of the proof but simply an example of the relation between ε and δ. To make the proof more concrete let $\varepsilon = 0.0007$. Then $\delta = \min\left(1, \frac{0.007}{5}\right) = 0.0014$. Thus $|x-2| < \delta - 0.0014$ translates to $1.9986 < x < 2.0014$. Thus $3.99440196 < x^2 < 4.00560196$ or $-0.00559804 < x^2 - 4 < 0.00560196$. Thus we have $|x^2 - 4| < 0.007$ for all those xs.

We now prove a theorem similar to one we proved shortly after making the definition of the limit of a sequence.

Theorem 7.4. Suppose that $f : A \to \mathbb{R}$, c is a limit point of A, and $\lim_{x\to c} f(x) = L$. Then there is an $\varepsilon > 0$ such that f is bounded on $N_\varepsilon(c) \cap A$.

Proof. To confuse the reader, in this theorem the role of ε will be played by the number 1 and the role of $\delta > 0$ will be played by ε. Since $\lim_{x\to c} f(x) = L$ we know that there is an $\varepsilon > 0$ (masquerading as δ) such that if $0 < |x-c| < \varepsilon$ and $x \in A$, then $|f(x) - L) < 1$. This last inequality may be replaced by $L - 1 < f(x) < L + 1$. Thus for all x in $N_\varepsilon(c) \cap A$ we have $f(x)$ bounded below by $L-1$ and bounded above by $L+1$ with the possible exception of $x = c$. Then $f(c) \neq L$ is a possibility if c is in the domain of f. Thus we can say that for all x in $N_\varepsilon(c) \cap A$ we have $\min\{f(c), L-1\} \leq f(x) \leq \max\{f(c), L+1\}$. Thus f is bounded on $N_\varepsilon(c) \cap A$. □

Another theorem that is similar to one we proved for sequential limits is:

Theorem 7.5 (Uniqueness of Limits of Functions). If $\lim_{x\to c} f(x) = L$ and $\lim_{x\to c} f(x) = M$, then $L = M$.

Proof. We assume that $L \neq M$ and let $\varepsilon = \frac{|L-M|}{2} > 0$. Assume that A is the domain of f and that c is a limit point of A. (All of this is implicit in the limit notation). Then there are positive numbers δ_1 and δ_2 such that if $0 < |x-c| < \delta_1$, then $|f(x) - L| < \frac{\varepsilon}{2}$ and if $0 < |x-c| < \delta_2$ then $|f(x) - M| < \frac{\varepsilon}{2}$. Now choose $\delta = \min\{\delta_1, \delta_2\}$. Then if $|x-c| < \delta$ we have

$$\begin{aligned} |L - M| &= |L - f(x) + f(x) - M| \\ &\leq |L - f(x)| + |f(x) - M| \\ &< \frac{\varepsilon}{2} + \frac{\varepsilon}{2} \\ &= \varepsilon = \frac{|L-M|}{2} \end{aligned}$$

This is a contradiction because no positive number (in this case $|L - M|$) can be less than one half itself. □

As we saw in the examples, proving a limit for a function as simple as $f(x) = x^2$ from the definition is not a trivial process. The next theorem gives us a way to compute limits without resorting to the definition at every turn.

Theorem 7.6. Let $\lim_{x\to c} f(x) = L$, $\lim_{x\to c} g(x) = M$, and k be a real number. Then:

(1) $\lim_{x\to c}(f(x) \pm g(x)) = L \pm M$,

(2) $\lim_{x\to c} kf(x) = kL$,

(3) $\lim_{x\to c}(f(x) \cdot k(x)) = L \cdot M$, and

(4) if $M \neq 0$, then $\lim_{x\to c}\left(\frac{f(x)}{g(x)}\right) = \frac{L}{M}$.

Proof. We prove the addition case ((1)) and the multiplication case ((3)).

For ((1)): Let $\varepsilon > 0$ be given. We assume that both functions have domain A and that c is a limit point of the common domain. Because the individual limits exist we can find $\delta_1 > 0$ and $\delta_2 > 0$ such that if $0 < |x - c| < \delta_1$ and $x \in A$, then $|f(x) - L| < \frac{\varepsilon}{2}$ and if $0 < |x - c| < \delta_2$ then $|g(x) - M| < \frac{\varepsilon}{2}$. Let $\delta = \min\{\delta_1, \delta_2\}$. Then if $0 < |x - c| < \delta$ and $x \in A$, then

$$\begin{aligned}
|(f(x) + g(x)) - (L + M)| &= |(f(x) - L) + (g(x) - M)| \\
&\leq |f(x) - L| + |g(x) - M| \\
&< \frac{\varepsilon}{2} + \frac{\varepsilon}{2} = \varepsilon.
\end{aligned}$$

The proof for subtraction is almost exactly the same.

For ((3)): The proof for multiplication is more involved. When we proved a similar result for sequences we required the boundedness of a convergent sequence. Theorem 7.4 gives us what we need in this proof.

Let $\varepsilon > 0$ be given. We begin with a short computation.

$$|f(x)g(x) - LM| - |f(x)g(x) - f(x)M + f(x)M - LM| \leq |f(x)||g(x) - M| + |f(x) - L||M|$$

By Theorem 7.4 there is a $\delta_1 < 0$ such that for all x satisfying $0 < |x - c| < \delta_1$ and $x \in A$, $|f(x)| \leq K$ for some real number K. (That is the function f is bounded on a certain set.)

Since $\lim_{x\to c} f(x) = L$, there exists a $\delta_2 > 0$ such that for all x satisfying $0 < |x - c| < \delta_2$ and $x \in A$,

$$|f(x) - L| < \frac{\varepsilon}{2(|M| + 1)}.$$

Since $\lim_{x\to c} g(x) = M$, there exists a $\delta_3 > 0$ such that for all x satisfying $0 < |x - c| < \delta_3$ and $x \in A$,

$$|g(x) - M| < \frac{\varepsilon}{2(|K| + 1)}.$$

Let $\delta = \min\{\delta_1, \delta_2, \delta_3\}$. Now assume that x satisfies $0 < |x - c| < \delta$ and $x \in A$. Using the short computation we started with we have:

$$\begin{aligned}
|f(x)g(x) - LM| &\leq |f(x)||g(x) - M| + |M||f(x) - L| \\
&< |K| \cdot \frac{\varepsilon}{2(|K| + 1)} + |M| \cdot \frac{\varepsilon}{2(|M| + 1)} \\
&< \frac{\varepsilon}{2} + \frac{\varepsilon}{2} = \varepsilon
\end{aligned}$$

□

Definition 7.7 (A Real Polynomial). A *real polynomial* or a *polynomial with real coefficients*, is a function of the following form:

$$f(x) = a_n x^n + a_{n-1}x^{n-1} + \cdots a_1 x + a_0,$$

where the a_k are real numbers and $n \in \mathbb{N} \cup \{0\}$.

Putting our Example 7.3 together with Theorem 7.6 we have :

Theorem 7.8. If $P(x)$ is a polynomial, then $\lim_{x\to c} P(x) = P(c)$ for any real number c. If $f(x) = \frac{P(x)}{Q(x)}$ is a quotient of polynomials and $Q(c) \neq 0$ then $\lim_{x\to c} f(x) = f(c)$.

Proof. We know that $\lim_{x\to c} x = c$ from Example 7.3.1.. Polynomials and quotients of polynomials are formed with real numbers and the four arithmetic operations, all of which fall under Theorem 7.6. □

A very important case of limits of quotients of polynomials occurs even when $Q(c) = 0$.

Example 7.9. Consider $f(x) = \frac{x^3-1}{x-1}$, which is undefined at $x = 1$. We can however compute $\lim_{x\to 1} f(x) = \lim_{x\to 1} \frac{x^3-1}{x-1}$. Factoring the numerator yields

$$\frac{x^3-1}{x-1} = \frac{(x-1)(x^2+x+1)}{x-1} = \left(\frac{x-1}{x-1}\right)(x^2+x+1).$$

Since $\frac{x-1}{x-1} = 1$ except when $x = 1$, we see that the function $f(x)$ is equal to the polynomial x^2+x+1 except at $x = 1$. Since our definition of $\lim_{x\to c} f(x)$ specifically ignores the value of f at $x = c$, we have

$$\lim_{x\to 1} \frac{x^3-1}{x-1} = \lim_{x\to 1} x^2+x+1 = 1^2+1+1 = 3.$$

Another way to look at this is that $\lim_{x\to c} \frac{x-c}{x-c} = 1$ since the quotient is equal to 1 for all x except $x = c$, where it is undefined.

We have seen that the definition of the limit of a function is more complex than that of the limit of a sequence, but there is an important connection between the two definitions that can be very useful.

Theorem 7.10. Let $f : A \to \mathbb{R}$ and let c be a limit point of A. Then $\lim_{x\to c} f(x) = L$ if and only if for every sequence $\{x_n\}$ contained in A satisfying $\lim_{n\to\infty} x_n = c$ and $x_n \neq c$ we have $\lim_{n\to\infty} f(x_n) = L$.

Proof. We assume that $\lim_{x\to c} f(x) = L$. Let $\{x_n\} \subset A$ be a sequence that converges to c and is never equal to c. Given $\varepsilon > 0$, there is a $\delta > 0$ such that for all x satisfying $0 < |x-c| < \delta$ and $x \in A$ it follows that $|f(x) - L| < \varepsilon$. Since $\{x_n\}$ converges to c, there is a natural number N such that if $n \geq N$ then $|x_n - c| < \delta$. Since c is not in the sequence we can say further that $0 < |x_n - c| < \delta$. Then by the existence of the limit we have $|f(x_n) - L| < \varepsilon$ which tells us that $\lim_{n\to\infty} f(x_n) = L$.

We can prove the opposite implication by contradiction, but leave the details to the reader. □

Definition 7.11 (Continuity). Let $f : A \to \mathbb{R}$ and $c \in A$. We say that f is *continuous at* c if for every $\varepsilon > 0$ there is a $\delta > 0$ such that if $|x-c| < \delta$ and $x \in A$, then $|f(x) - f(c)| < \varepsilon$. We say that f is *continuous on the set* A if it is continuous at all points c in A.

Remark 7.12. Again the definition of continuity is slightly different from that given in most Calculus texts. Often the given definition is that f is continuous at c if $\lim_{x\to c} f(x) = f(c)$. If c is an isolated point of A then f is automatically continuous there under our definition but $\lim_{x\to c} f(x)$ but does not exist and hence f would not be continuous using the calculus definition. Notice also that the definition of continuity at c differs only slightly from the definition of limit. The point c must be in domain A but does not have to be a limit point of A. The inequality $|x - c| < \delta$ does not require that x be different from c. Finally L is replaced by $f(c)$.

What does continuity mean? We have defined continuity at a point of the domain of a function. Essentially f is continuous at c if whenever x in the domain of f is close to c then $f(x)$ is close to $f(c)$. Suppose there are no xs close to c, that is, suppose c is an isolated point of the domain. Then f must be continuous at c.

Theorem 7.13. Let $f : A \to \mathbb{R}$ and let $c \in A$. If c is an isolated point of A, then f is continuous at c. If c is a limit point of A, then f is continuous at c if and only if $\lim_{x\to c} f(x) = f(c)$.

Proof. Recall that c is an isolated point of A if it is not a limit point of A. Thus if c is an isolated point of A then there is a $\delta > 0$ such that $N_\delta(c) \cap A = \{c\}$. If $|x - c| < \delta$ and $x \in A$, then $x = c$. Hence $|f(x) - f(c)| = 0 < \varepsilon$ for all $\varepsilon > 0$. Thus f is continuous at c.

Now assume that c is a limit point of A. If f is continuous at c then for a given $\varepsilon > 0$ there is a $\delta > 0$ such that for all x satisfying $|x - c| < \delta$ and $x \in A$ it follows that $|f(x) - f(c)| < \varepsilon$. This implies that $\lim_{x\to c} f(x) = f(c)$. □

Theorem 7.14. All polynomials are continuous at every real number and all quotients of polynomials are continuous at every real number different from a root of the denominator polynomial.

Proof. This follows immediately from Theorem 7.8. □

Example 7.15. Let us prove that $f(x) = x^2$ is continuous for all x directly from the definition. Let c be a real number. We show that f is continuous at c.

Suppose $c = 0$. Given $\varepsilon > 0$, let $\delta = \min\{\varepsilon, 1\}$. If $|x - 0| = |x| < \delta$ then $|x^2 - 0| = |x^2| < |x| < \varepsilon$. Now suppose $c \neq 0$. Given $\varepsilon > 0$, let $\delta = \min\left\{1, \frac{\varepsilon}{2|c|+1}\right\}$. If $|x - c| < 1$ then $c - 1 < x < c + 1$ and hence $|x + c| < 2|c| + 1$. If $|x - c| < \delta$ we have $|x^2 - c^2| = |x - c||x + c| < \left(\frac{\varepsilon}{2|c|+1}\right)(2|c| + 1) = \varepsilon$.

A very useful fact is that continuity is preserved under composition of functions.

Theorem 7.16. Let $A, B \subset \mathbb{R}$ and let $c \in A$. Let $f : A \to \mathbb{R}$, $g : B \to \mathbb{R}$, and assume that $f(A) \subset B$. Suppose that f is continuous at c and that g is continuous at $f(c)$. Then $g \circ f$ is continuous at c.

Proof. Let $\varepsilon > 0$ be given. Since g is continuous at $f(c)$, there is a $\beta > 0$ such that if $|y - f(c)| < \beta$ and $y \in B$, then $|g(y) - g(f(c))| < \varepsilon$. Since f is continuous at c, there is a $\delta > 0$ such that if $|x - c| < \delta$ and $x \in A$, then $|f(x) - f(c)| < \beta$. Combining the two sets of inequalities we have that if $|x - c| < \delta$ and $x \in A$, then $|g(f(x)) - g(f(c))| < \varepsilon$. □

A form of Theorem 7.10 is applicable to continuous functions.

Theorem 7.17. Let $f : A \to \mathbb{R}$ be continuous on $A \subset \mathbb{R}$ and let $\lim_{x\to\infty} x_n = x$ for some sequence $\{x_n\}$ contained in A. Assume that x is in A. Then $\lim_{n\to\infty} f(x_n) = f(x)$.

Theorem 7.18. Suppose $f : K \to \mathbb{R}$ is continuous and $K \subset \mathbb{R}$ is compact. Then $f(K)$ is also compact.

Proof. Continuity does not preserve boundedness. (For instance, $f(x) = \frac{1}{x}$ is continuous on $(0, 1)$ but its image is not bounded.) Thus it will be difficult to prove this result using the Heine-Borel Theorem (Theorem 6.25). We need to go back to the definition of compactness. Let $\{y_n\}$ be a sequence contained in $f(K)$. For each n we choose an $x_n \in K$ such that $f(x_n) = y_n$. The sequence $\{x_n\}$ has a subsequence that converges to a point of K since K is compact by assumption. We will call the subsequence $\{x_n\}$, the same as the original sequence. Let $y_n = f(x_n)$. This defines a subsequence of the original sequence in $f(K)$. Since $\lim_{x\to\infty} x_n = x$ with $x \in K$ and f is continuous at x by assumption, we know that $\lim_{x\to\infty} f(x_n) = f(x)$ by the previous theorem. Thus $f(x)$ is in $f(K)$ and is the limit of the subsequence of the original sequence. □

Theorem 7.19. Let K be a non-empty compact set. Then K contains its least upper bound and its greatest lower bound.

Proof. Since K is compact it is closed and bounded. Since K is non-empty, then K has both a least upper bound and a greatest lower bound by the Completeness Axiom (Axiom 6). Let d be the least upper bound of K. We will prove that the least upper bound of a set is either in the set or is a limit point of the set. Suppose that a is the least upper bound for some set A and assume that a is not an element of A. Then by the ε-criterion for least upper bounds (Theorem 3.5), for each $\varepsilon > 0$ there is an x in A such that $a - \varepsilon < x$. Since a is not in A, it is not equal to x and hence there is an element of A different from a in every ε-neighborhood of a. Therefore a is a limit point of A. Since the set K is closed it contains all its limits points and hence $d \in K$. The proof for the greatest lower bound is similar. □

The first application of these two theorems is a very important result from Calculus.

Theorem 7.20 (Extreme Value Theorem)**.** Let $f : [a, b] \to \mathbb{R}$ be continuous. Then there exist points c and d in $[a, b]$ such that $f(c) \le f(x) \le f(d)$ for all $x \in [a, b]$. The point c is called *an absolute, or global, minimum point* of f on $[a, b]$ and d is called *an absolute, or global, maximum point* of f on $[a, b]$.

Proof. We will just show the existence of the point d. Since f is continuous we know that $f([a, b])$ is compact. Thus it contains its least upper bound by the previous theorem. Call it y. Since $y \in f([a, b])$ we are done, since $y = f(d)$ for some $d \in [a, b]$. □

Definition 7.21 (Uniform Continuity)**.** A function f is *uniformly continuous* on a set A if for every $\varepsilon > 0$ there is a $\delta > 0$ such that if $|x - y| < \delta$ and $x, y \in A$ then $|f(x) - f(y)| < \varepsilon$.

Notice that this differs from continuity in that the choice of δ does not depend on the point in the domain.

Example 7.22. The function $f : [0, 3] \to \mathbb{R}$ defined by $f(x) = x^2$ is uniformly continuous on $[0, 3]$. Let $\varepsilon > 0$ be given and let x and y be any points in $[0, 3]$. Then

$$\begin{aligned} |f(x) - f(y)| &= |x^2 - y^2| \\ &= |x - y||x + y| \\ &\le |x - y|\,(|x| + |y|) \\ &\le 6|x - y|. \end{aligned}$$

Let $\delta = \frac{\varepsilon}{6}$ and we have uniform continuity.

Theorem 7.23 (Criterion for non-uniform continuity)**.** Let $f : A \to \mathbb{R}$. There are two sequences in A, $\{x_n\}$ and $\{y_n\}$, and an $\varepsilon_0 > 0$ such that $\lim_{n\to\infty} |x_n - y_n| = 0$ and for all n, $|f(x_n) - f(y_n)| \geq \varepsilon_0$ if and only if f is not uniformly continuous on A.

Proof. The theorem follows from negating the definition of uniform continuity. Given any $\delta > 0$ we can find an n such that $|x_n - y_n| < \delta$ and $|f(x_n) - f(y_n)| \geq \varepsilon_0$. This violates uniform continuity. □

Theorem 7.24. If $f : A \to \mathbb{R}$ is continuous and A is compact then f is uniformly continuous on A.

Proof. We assume that f is not uniformly continuous and take two sequences, $\{x_n\}$ and $\{y_n\}$ in A, and an $\varepsilon_0 > 0$ such that $\lim_{n\to\infty} |x_n - y_n| = 0$ and $|f(x_n) - f(y_n)| \geq \varepsilon_0$ for all n. Since A is compact there is a subsequence $\{x_{i_n}\}$ of $\{x_n\}$ that converges to x in A. Consider the corresponding subsequence $\{y_{i_n}\}$ of $\{y_n\}$. Since $\lim_{n\to\infty} |x_n - y_n| = 0$ we know that $\lim_{n\to\infty} x_{i_n} - y_{i_n} = 0$ also and hence that $\lim_{n\to\infty} y_{i_n} = x$ By continuity since $\lim_{n\to\infty} x_{i_n} - y_{i_n} = 0$ then $\lim_{n\to\infty} f(x_{i_n}) - f(y_{i_n}) = 0$ which violates the ε_0 condition. □

Example 7.25. The function $f : [0,\infty) \to \mathbb{R}$ defined by $f(x) = x^2$ is continuous but not uniformly continuous. Consider the sequences $\{x_n\} = \{n\}$ and $\{y_n\} = \{n + \frac{1}{n}\}$. Then $\lim_{n\to\infty} |x_n - y_n| = \lim_{n\to\infty} \frac{1}{n} = 0$ but $|f(x_n) - f(y_n)| = \left|n^2 - \left(n + \frac{1}{n}\right)^2\right| = \left|n^2 - n^2 - 2 - \frac{1}{n^2}\right| = 2 + \frac{1}{n^2} \geq 2$. Choosing $\varepsilon_0 = 2$ gives us non-uniform continuity.

Uniform continuity will become important when we deal later with the Riemann Integral.

We now prove another important theorem from Calculus the Intermediate Value Theorem. The proof requires the notion of a connected set. For the real numbers, a connected set is simply an interval. The idea is also important in higher dimensions and we give the definition that is used in this more general setting.

Definition 7.26 (Connectedness)**.** Let A and B be non-empty subsets of the real numbers. We say that A and B are *separated* if both $\overline{A} \cap B$ and $A \cap \overline{B}$ are empty. A set C is *disconnected* if C is the union of two separated sets. We say C is *connected* if C is not disconnected.

This is a complicated definition, saying that C is connected if it fails to have some non-obvious property.

Example 7.27. Let $A = (0,1)$, $B = (1,2)$, and $C = [1,2)$. Then A and B are separated (because $\overline{A} \cap B = [0,1] \cap (1,2) = \emptyset$ and $A \cap \overline{B} = (0,1) \cap [1,2] = \emptyset$) while A and C are not (because $\overline{A} \cap C = [0,1] \cap [1,2) = 1 \neq \emptyset$).

Theorem 7.28. If $f : A \to \mathbb{R}$ is continuous and B is a connected subset of A then $f(B)$ is connected.

Proof. This theorem tells us that a continuous function cannot rip a connected set into two separated pieces. We proceed by contradiction. Assume that B is connected but $f(B)$ is not. Then we can write $f(B) = U \cup V$ where U and V are separated sets. Consider the sets $X = f^{-1}(U) \cap B$ and $Y = f^{-1}(V) \cap B$. Then $B = U \cup V$. Now suppose that $\overline{X} \cap Y \neq \emptyset$. Let $a \in \overline{X} \cap Y$. Then a cannot be in both X and Y. If it were then we would have $f(a) \in U$ and $f(a) \in V$ but U and V are disjoint sets. Thus a must be an element of Y and a limit point of X. We show that $f(a)$ is a limit point of U. Let $\varepsilon > 0$ be given. Then there is a $\delta > 0$ such that if $|x - a| < \delta$ and $x \in A$ then $|f(x) - f(a)| < \varepsilon$. Since a is a limit point of X there must be a y in $N_\delta(a) \cap X$ that is different from a. Since a is in Y but not in X (it is a limit point of X but not in X), $f(x) \neq f(a)$ because

$f(y) \in f(X) = U$ and $f(a) \in f(Y) = V$. Thus for any $\varepsilon > 0$ we can find a point of $f(X)$ that is in $N_\varepsilon(f(a)) \cap U$ and different from $f(a)$. Thus $f(a)$ is a limit point of U and $\overline{U} \cap V \neq \emptyset$. Therefore U and V are not separated so $f(B)$ is connected. □

Theorem 7.29. A set C of real numbers is connected if and only if whenever a and b are points of C with $a < b$ and x is any real number satisfying $a < x < b$ then x is a point of C.

Proof. Let C be a connected set and let $a < b$ for two points in C. Let x satisfy $a < x < b$. We must show that $x \in C$. Suppose it is not. Consider the sets $U = C \cap (-\infty, x)$ and $V = C \cap (x, \infty)$. Then $C = U \cap V$, $\overline{U} \cap V = \emptyset$, and $U \cap \overline{V} = \emptyset$. Thus C is the union of separated sets and is not connected.

Now for the opposite implication. Suppose that C satisfies the property that all xs between any two elements of C are also in C. We show that C is connected. Suppose that C is disconnected and that $C = U \cup V$ where U and V are separated. Let $a \in U$ and $b \in V$ satisfy $a < b$. Consider $X = \{x \in V \mid a < x\}$. The set X is non-empty (it contains b) and X is bounded below (by a). Thus it has a greatest lower bound. Call it y. Then $a < y < b$ or $y = a$ or $y = b$. In any case $y \in C$. Thus $y \in U$ or $y \in V$, but not both. If $y \in U$ then as the greatest lower bound of X it is a limit point of V and hence $U \cap \overline{V} \neq \emptyset$. This violates our assumption. Thus $y \in V$. But by our condition every z between a and y is in C. An element z cannot be in V. Thus it is in U. Thus y is a limit point of U and hence $\overline{U} \cap V \neq \emptyset$. This is the final contradiction. □

Now we have the very important:

Theorem 7.30 (Intermediate Value Theorem)**.** Let $f : [a, b] \to \mathbb{R}$ be continuous and let d be a real number satisfying either $f(a) < d < f(b)$ or $f(b) < d < f(a)$. Then there is a real number c in $[a, b]$ such that $f(c) = d$.

Proof. Assume $f(a) < d < f(b)$. We know that $[a, b]$ is connected by Theorem 7.29, and hence that $f([a, b])$ is connected by Theorem 7.28. Using Theorem 7.29 again we find that $d \in f([a, b])$ and hence that $d = f(c)$ for some c in $[a, b]$. □

7.1 Exercises

Exercise 7.1. Prove that $\lim_{x\to 3} 5x - 6 = 9$.

Exercise 7.2. Prove that $\lim_{x\to -3} x^2 = 9$.

Exercise 7.3. Prove that $\lim_{x\to 3} \frac{1}{x} = \frac{1}{3}$.

Exercise 7.4. Prove that $\lim_{x\to 0} |x| = 0$.

Exercise 7.5. Prove that $\lim_{x\to 3} x^3 + 1 = 28$.

Exercise 7.6. A function $f(x)$ is said to be *bounded on a set* $A \subset \mathbb{R}$ if there is a positive real number M such that for all a in A, $|f(a)| \leq M$. Suppose that A is a set of real numbers, c is a limit point of A, f is bounded on A, $g(x)$ is defined on A, and $\lim_{x\to c} g(x) = 0$. Prove that $\lim_{x\to c} f(x)g(x) = 0$.

Exercise 7.7. Let $f : \mathbb{R} \to \mathbb{R}$ be continuous on $\mathbb{R}$. Let $K = \{x \in \mathbb{R} \mid f(x) = 0\}$. Prove that K is closed. (Hint: Begin with "Let c be a limit point of K." What do you have to prove about $f(c)$?)

Exercise 7.8. Let $f : \mathbb{R} \to \mathbb{R}$ and let c be a real number, $0 < c < 1$, such that $|f(x)-f(y)| \leq c|x-y|$ for all x and y in $\mathbb{R}$. Show that f is continuous on $\mathbb{R}$.

Exercise 7.9. Let $f : [1, 2] \to \mathbb{R}$ be defined by $f(x) = \frac{1}{x}$. Show that f is uniformly continuous on $[1, 2]$ by finding $\delta > 0$ that satisfies the definition fo ra given $\varepsilon > 0$. More specifically find δ as a function of ε.

Exercise 7.10. Let $f(x)$ be defined as below. Show that f is not continuous at $x = 0$.

$$f(x) = \begin{cases} \sin\left(\frac{1}{x}\right) & \text{if } x \neq 0 \\ 0 & \text{if } x = 0 \end{cases}$$

Exercise 7.11. Let $f(x) = \frac{1}{x}$ on $(0, \infty)$. Show that f is not uniformly continuous on that interval.

Exercise 7.12. Suppose that $f : A \to \mathbb{R}$ is uniformly continuous and that $\{x_n\}$ is a Cauchy sequence in $A \subset \mathbb{R}$. Prove that $\{f(x_n)\}$ is a Cauchy sequence.

Exercise 7.13. Let $f(x) = x^2 - 2x$. Let d be a real number satisfying $-1 < d < 15$. Explicitly find a c satisfying $1 < c < 5$ such that $f(c) = d$.

Exercise 7.14. Let $f : \mathbb{R} \to \mathbb{R}$ be defined by $f(x) = 0$ if x is a rational number and $f(x) = 1$ if x is an irrational number. Show that $\lim_{x\to c} f(x)$ does not exist for any real number c.

Exercise 7.15. Let $f : \mathbb{R} \to \mathbb{R}$ be defined by $f(x) = 0$ if x is a rational number and $f(x) = x$ if x is an irrational number. Show that $\lim_{x\to 0} f(x) = 0$ and that $\lim_{x\to c} f(x)$ does not exist for any real number c different from 0.

Exercise 7.16. Let $f : \mathbb{R} \to \mathbb{R}$ and assume that $\lim_{x\to 0} f(x) = 3$. Show that there is an $\varepsilon > 0$ such that if $0 < |x - 0| < \varepsilon$ then $f(x) > 1$.

Exercise 7.17. Suppose that A is a non-empty set of real numbers that is bounded above. Let a be the least upper bound of A and assume that a is not in A. Prove that a is a limit point of A.

Chapter 8

The Derivative

We begin with the definition of the derivative of a function at a point in its domain. Note that the domain is required to be an interval. We are not defining the derivative at an isolated point of the domain.

Definition 8.1 (Differentiable)**.** Let A be an interval. Let $f : A \to \mathbb{R}$ and let $a \in A$. The *derivative of* f *at* a is $\lim_{x\to a} \frac{f(x)-f(a)}{x-a}$ (if the limit exists). If the limit exists we denote it by $f'(a) = \lim_{x\to a} \frac{f(x)-f(a)}{x-a}$ and say that f is *differentiable at* a. If the limit does not exist then f *is not differentiable at* a.

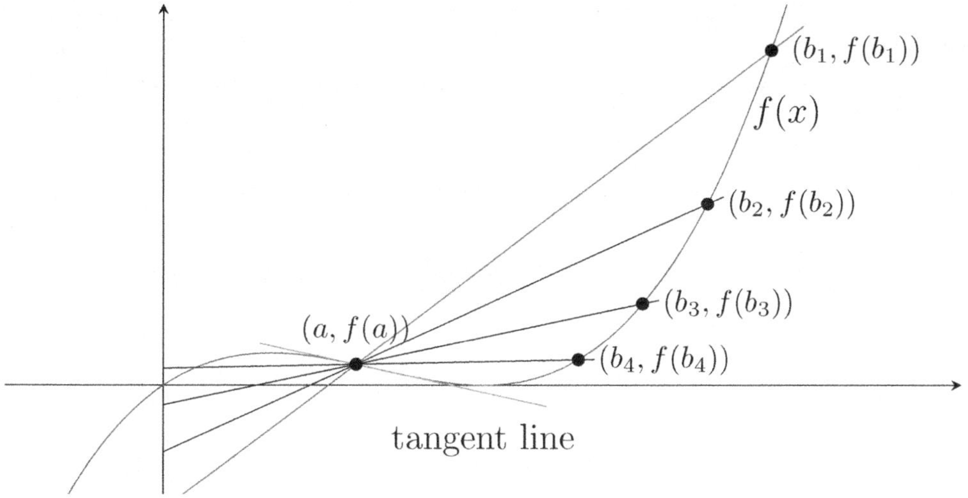

Figure 8.1

Remark 8.2. Several secant lines through the point $(a, f(a))$ are shown in Figure 8.1; the second point on $f(x)$ for each is marked $(b_i, f(b_i))$ and as i increases, b_i approaches a. The line tangent to $f(x)$ at $x = a$ is shown in red, and the figure implies that as b_i approaches a, the corresponding secant lines approach the tangent line. Thus the limit of the slopes of the secant lines approaches the slope of the tangent line.

Remark 8.3. Letting $h = x - a$ in the above definition yields $f'(a) = \lim_{h\to 0} \frac{f(a+h)-f(a)}{h}$. This is often a very useful form for computing the derivative.

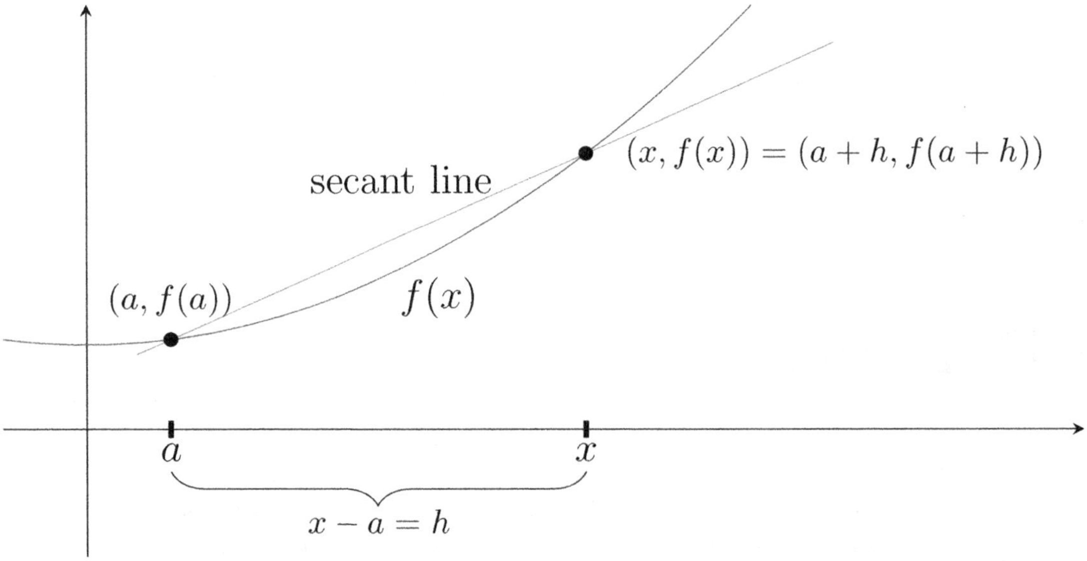

Figure 8.2

Remark 8.4. Figure 8.2 depicts the secant line to the graph of $f(x)$ through the points $(a, f(a))$ and $(x, f(x))$ on the graph. It also shows x replaced by $a+h$. As x approaches a, or equivalently as h approaches 0, this secant line bends towards the tangent line at a. Thus intuitively the derivative measures the slope of the tangent line since it is the limit of the slopes of the secant lines.

Definition 8.5 (Derivative of a Function)**.** Let $f : A \to \mathbb{R}$. The function $f'(x) = \lim_{h\to 0} \frac{f(x+h)-f(x)}{h}$ is called the *derivative of* $f(x)$. If $f'(x)$ exists for all $x \in A$ then we say that f is *differentiable on* A.

Example 8.6.

1. Let $f(x) = x^2$ for all real numbers x. We compute $f'(a)$ using both forms of the definition.

$$f'(a) = \lim_{x\to a} \frac{f(x) - f(a)}{x - a} = \lim_{x\to a} \frac{x^2 - a^2}{x - a} = \lim_{x\to a} \left(\frac{x-a}{x-a}\right)(x + a) = 2a$$

$$\begin{aligned} f'(a) = \lim_{j\to 0} \frac{f(a+h) - f(a)}{h} &= \lim_{h\to 0} \frac{(a+h)^2 - a^2}{h} \\ &= \lim_{h\to 0} \frac{a^2 + 2ah + h^2 - a^2}{h} \\ &= \lim_{h\to 0} \left(\frac{h}{h}\right)(2a + h) \\ &= 2a \end{aligned}$$

2. Let $f(x) = |x|$ for all real numbers x. Then $f'(0)$ does not exist.

$$f'(0) = \lim_{h\to 0} \frac{|0 + h| - |0|}{h} = \lim_{h\to 0} \frac{|h|}{h}$$

This limit does not exist since we can write $\frac{|h|}{h} = \begin{cases} -1 & h < 0 \\ 1 & h > 0 \end{cases}$.

Theorem 8.7. If $f : A \to \mathbb{R}$ is differentiable at $a \in A$ then f is continuous at $a \in A$.

Proof. Since a is a limit point of A, it suffices to show that $\lim_{x\to a} f(x) = f(a)$ or what is equivalent $\lim_{x\to a} f(x) - f(a) = 0$. We know that $\lim_{x\to a} \frac{f(x)-f(a)}{x-a} = f'(a)$ and that $\lim_{x\to a} x - a = 0$. The limit of the product equals the product of the limits. Thus

$$\lim_{x\to a} \left(\frac{f(x) - f(a)}{x - a}\right)(x - a) = f'(a) \cdot 0 = 0 = \lim_{x\to a} f(x) - f(a). \qquad \square$$

Now a theorem that contains the product rule, quotient rule, and the power rule.

Theorem 8.8. Suppose that f and g are differentiable at a and that r is a real number. Then all of the following rules are true for f and g at a.

(1) $(rf)'(a) = rf'(a)$

(2) $(f + g)'(a) = f'(a) + g'(a)$

(3) $(fg)'(a) = f'(a)g(a) + f(a)g'(a)$

(4) If $g(a) \neq 0$, then $\left(\frac{f}{g}\right)'(a) = \frac{f'(a)g(a)-f(a)g'(a)}{(g(a))^2}$.

(5) If $n \in \mathbb{Z}$ and $f(x) = x^n$ then $f'(x) = nx^{n-1}$.

Proof. We will prove (3), the product rule and (5), the power rule.

For (3):

$$\begin{aligned}(fg)'(a) &= \lim_{x\to a} \frac{f(x)g(x) - f(a)g(a)}{x - a} = \lim_{x\to a} \frac{f(x)g(x) - f(a)g(x) + f(a)g(x) - f(a)g(a)}{x - a} \\ &= \lim_{x\to a} \left(\frac{f(x) - f(a)}{x - a}\right) g(x) + f(a)\left(\frac{g(x) - g(a)}{x - a}\right) \\ &= f'(a)g(a) + f(a)g'(a).\end{aligned}$$

The only subtlety is that $\lim_{x\to a} g(x) = g(a)$ because g is continuous at a. (Why?)

For (5): Let $n = 0$. Then $f(x) = 1$ for all x and hence the derivative is the zero function as the theorem requires. We prove the result for $n \in \mathbb{N}$ by induction.

The base case is when $n = 1$. A very simple limit calculation shows that if $f(x) = x$, then $f'(x) = 1$. For the induction step, assume that if $f(x) = x^k$, then $f'(x) = kx^{k-1}$. Now let $g(x) = x^{k+1} = x \cdot x^k$. By the product rule $g'(x) = 1 \cdot x^k + x \cdot kx^{k-1} = x^k + kx^k = (k + 1)x^k$, as required.

Now for the rest of the integers. Assume that $n < 0$. Let $m = -n$. Then $f(x) = x^n = x^{-m} = \frac{1}{x^m}$. We apply the quotient rule to f in this form.

$$f'(x) = \frac{0 \cdot x^m - 1 \cdot mx^{m-1}}{(x^m)^2} = \frac{-mx^{m-1}}{x^{2m}} = \frac{-m}{x^{m+1}} = \frac{n}{x^{-n+1}} = nx^{n-1} \qquad \square$$

Theorem 8.9 (The Chain Rule). Let $f : A \to \mathbb{R}$, $g : B \to \mathbb{R}$, and $f(a) \subset B$. Suppose that f is differentiable at c and that g is differentiable at $f(c)$. Then $g \circ f$ is differentiable at c and $(g \circ f)'(c) = g'(f(c))f'(c)$.

Proof. We first define two new functions closely related to f and g:

$$\hat{f}(x) = \begin{cases} \frac{f(x)-f(c)}{x-c} & x \neq c \\ f'(c) & x = c \end{cases} \text{ and } \hat{g}(y) = \begin{cases} \frac{g(y)-g(f(c))}{y-f(c)} & y \neq f(c) \\ g'(f(c)) & y = f(c) \end{cases}.$$

Since both f and g are differentiable at the appropriate points these functions are clearly continuous at $x = c$ and at $y = f(c)$. Consider the product:

$$\hat{f}(x)\hat{g}(f(x)) = \begin{cases} \left(\frac{f(x)-f(c)}{x-c}\right)\left(\frac{g(f(x))-g(f(c))}{f(x)-f(c)}\right) & x \neq c \\ f'(c)g'(f(c)) & x = c \end{cases}.$$

If we could claim that $f(x) \neq f(c)$ in a neighborhood of c, we would be done since that would reduce $\hat{f}(x)\hat{g}(f(x))$ to

$$\hat{f}(x)\hat{g}(f(x)) = \begin{cases} \left(\frac{g(f(x))-g(f(c))}{x-c}\right) & x \neq c \\ f'(c)g'(f(c)) & x = c \end{cases}.$$

Since both $\hat{f}$ and $\hat{g}$ are continuous at c, so is the function $\hat{f}(x)\hat{g}(f(x))$ and hence

$$\lim_{x\to c} \hat{f}(x)\hat{g}(f(x)) = \lim_{x\to c} \frac{g(f(x)) - g(f(c))}{x - c} = f'(c)g'(f(c)) = (g \circ f)'(c).$$

However it is possible that in every neighborhood of c there is an x in A, different from c satisfying $f(x) = f(c)$. Let $\{x_n\}$ be a sequence of such points that converges to c. Then

$$\lim_{n\to\infty} \frac{f(x_n) - f(c)}{x_n - c} = \lim_{n\to\infty} \frac{0}{x_n - c} = 0.$$

But that limit is $f'(c)$. Now consider

$$\lim_{n\to\infty} g(f(x_n)) - g(f(c)) = g(f(c)) - g(f(c)) = 0.$$

This gives us a sequence $\{y_n\}$ in B such that $(g \circ f)'(c) = \lim_{n\to\infty} \frac{g(y_n)-g(f(c))}{y_n - f(c)} = 0$. Thus $0 = (g \circ f)'(c) = 0 \cdot g'(f(c))$. □

Definition 8.10 (Local Maxima and Minima)**.** Let f be defined on the open interval (a, b). A point c of the interval is a *local maximum point of* f if there is an open interval I contained in (a, b) such that c is in I and $f(x) \leq f(c)$ for all x in I. The number $f(c)$ is called a *local maximum value of* f. A similar definition holds for local minimum. The word *relative* is often used instead of the word *local.*

Definition 8.11 (Absolute Maxima and Minima)**.** Let f be defined on a set K. The function f has an *absolute maximum point at* c *in* K if $f(c) \geq f(x)$ for all x in the set K. In this case $f(c)$ is called the *absolute maximum value of* f *on* K. A similar definition holds for absolute minimum. The word *global* is often used instead of *absolute.*

Definition 8.12 (Monotonic Functions)**.** Let f be defined on an interval I. The function f is said to be *increasing on* I if whenever $x_1 < x_2$ for points in I then $f(x_1) < f(x_2)$. Reversing the inequality to $f(x_1) > f(x_2)$ yields the definition of a *decreasing function.* If the strict inequalities are replaced by greater than or equal or by less than or equal we say that f is either *non-decreasing* or *non-increasing*, respectively. A function fitting any one of these definitions is referred to as *monotonic*.

Lemma 8.13. Suppose that f is differentiable at c and that $f'(c) > 0$. Then there is a $\delta > 0$ such that for all h satisfying $0 < h < \delta$ it follows that $f(c-h) < f(c) < f(c+h)$. (If $f'(c) < 0$ we have $f(c-h) > f(c) > f(c+h)$.)

Proof. Assume that $f'(c) = \lim_{h\to 0} \frac{f(c+h)-f(c)}{h} > 0$. Since the limit is positive we know that there is a $\delta > 0$ such that $\frac{f(c+h)-f(c)}{h} > 0$ for all $0 < |h| < \delta$. If $h > 0$ then the numerator $f(c+h) - f(c) > 0$ or $f(c) < f(c+h)$. We have half of our inequality. Now assume that $h > 0$ and let $t = -h$.

$$\frac{f(c+t)-f(c)}{t} = \frac{f(c-h)-f(c)}{-h} > 0$$

Since the denominator is negative, so is the numerator and hence $f(ch) < f(c)$. □

Theorem 8.14 (Fermat)**.** Let $f : (a, b) \to \mathbb{R}$ with $c \in (a, b)$. Suppose that f has a local maximum or local minimum point at c and that f is differentiable at c. Then $f'(c) = 0$.

Proof. There are four possibilities for $f'(c)$: it is positive, it is negative, it does not exist, or it is zero. If $f'(c) > 0$ then Lemma 8.13 says that c is not a local maximum or minimum point of f. The same holds for the case that $f'(c) < 0$. Since we are assuming that f is differentiable at c the only remaining possibility is that $f'(c) = 0$. □

Theorem 8.15 (Rolle's Theorem)**.** Let f be continuous on $[a, b]$, differentiable on (a, b), and satisfy $f(a) = f(b)$. Then for some c in (a, b), $f'(c) = 0$.

Proof. By the Extreme Value Theorem (Theorem 7.20), f has both an absolute maximum point and an absolute minimum point in the interval $[a, b]$. Suppose neither of these points occurs in the interior of the interval. Then a and b are the absolute maximum and minimum points of f and hence f must be constant on the interval. Let c be the midpoint of the interval. Then $f'(c) = 0$.

Now assume that one of either the maximum or minimum points occurs at an interior point c of the interval. Then c will be a local maximum or local minimum point and hence $f'(c) = 0$ by Fermat's Theorem (Theorem 8.14). □

The next theorem, the Mean Value Theorem, is a very important result. It is the basis for the study of ordinary differential equations and various theoriest of approximation of functions via Taylor polynomials. Basically it takes our knowledge of the derivative of a function at individual points and extends it to knowledge about the function over an open interval. The proof of this theorem is essentially the proof of Rolle's Theorem (Theorem 8.15) with a secant line replacing the x-axis.

Theorem 8.16 (The Mean Value Theorem)**.** Let f be continuous on $[a, b]$ and differentiable on (a, b). Then there is a c in (a, b) satisfying$f'(c) = \frac{f(b)-f(a)}{b-a}$.

Proof. The Mean Value Theorem is essentially Rolle's Theorem tilted. The quantity $\frac{f(b)-f(a)}{b-a}$ is the slope of the secant line joining $(a, f(a))$ and $(b, f(b))$. Let $L(x)$ be the equation of that secant line. The equation for L is:

$$y = L(x) = \left(\frac{f(b)-f(a)}{b-a}\right)(x-a) + f(a).$$

Consider the function $H : [a,b] \to \mathbb{R}$ defined by $H(x) = f(x) - L(x)$. The function H is clearly continuous on the closed interval and $H'(x) = f'(x) - \frac{f(b)-f(a)}{b-a}$ by a simple calculation. Computing values of H we have:

$$H(a) = f(a) - \left(\frac{f(b)-f(a)}{b-a}\right)(a-a) - f(a) = 0 \text{ and}$$
$$H(b) = f(b) - \left(\frac{f(b)-f(a)}{b-a}\right)(b-a) - f(a) = 0.$$

Thus the function H satisfies the hypotheses of Rolle's Theorem. There is a c in the interior of the interval at which $H'(c) = 0$. Thus $H'(c) = f'(c) - \frac{f(b)-f(a)}{b-a} = 0$, or $f'(c) = \frac{f(b)-f(a)}{b-a}$. □

The application we make of the Mean Value Theorem (Theorem 8.16) is an important result used in first year calculus, relating a monotonic function and the sign of its derivative.

Theorem 8.17. Let I be an open interval.

(1) If $f'(x) > 0$ for all $x \in I$, then f is increasing on I.

(2) If $f'(x) < 0$ for all $x \in I$, then f is decreasing on I.

(3) If $f'(x) = 0$ for $x \in I$, then f is constant on I.

Proof. We prove (1) only. The other two parts are very similar. Let $x_1, x_2 \in I$ and $x_1 < x_2$. Since f is differentiable on the open interval I it is continuous on I. We apply the Mean Value Theorem to f over the interval $[x_1, x_2]$. There is a c satisfying $x_1 < c < x_2$ such that $f'(c) = \frac{f(x_2)-f(x_1)}{x_2-x_1}$. Since the left hand side is assumed to be positive by hypothesis and the denominator of the right hand side is positive by assumption it follows that $f(x_1) < f(x_2)$ and hence that f is increasing on I. □

The next theorem looks like the Mean Value Theorem (Theorem 8.16) from the standpoint of two functions, f and g, and shows that there is a single point, c, in the given interval that simultaneously satisfied the Mean Value Theorem for both these functions.

Theorem 8.18 (The Generalized Mean Value Theorem)**.** Let f and g be continuous on $[a,b]$ and differentiable on (a,b). Then there is a c in (a,b) such that $f'(c)(g(b) - g(a)) = g'(c)(f(b) - f(a))$.

For a proof, see Exercise 8.15. The Generalized Mean Value Theorem (Theorem 8.18) is the means by which we can proof the following version of L'Hospital's Rule (Theorem 8.19).

Theorem 8.19 (L'Hospital's Rule)**.** Let $f, g : A \to \mathbb{R}$ and let c be a limit point of A. Further suppose that $\lim_{x\to c} f(x) = \lim_{x\to c} g(x) = 0$ and that both $\lim_{x\to c} f'(x)$ and $\lim_{x\to c} g'(x)$ exist. Then $\lim_{x\to c} \frac{f(x)}{g(x)} = \lim_{x\to c} \frac{f'(x)}{g'(x)}$.

A final, though less direct, application of the Mean Value Theorem (Theorem 8.16) is called Daboux's Theorem, or the Intermediate Value Theorem for Derivatives (Theorem 8.20). It leads us to deeper insights about functions differentiable on open intervals.

Theorem 8.20 (Darboux's Theorem, Intermediate Value Theorem for Derivatives)**.** Let $f : [a,b] \to \mathbb{R}$ be differentiable on $[a,b]$. Assume that d is a real number satisfying either $f'(a) < d < f'(b)$ or $f'(b) < d < f'(a)$. Then there is a $c \in [a,b]$ such that $f'(c) = d$.

Proof. We cannot use the Intermediate Value Theorem (Theorem 7.30) because we know of an example of a function that is differentiable on an interval but whose derivative is not continuous there.

Assume that $f'(a) < d < f'(b)$. Define $g(x) = f(x) - dx$. Then g is differentiable on $[a, b]$ and $g'(x) = f'(x) - d$. We want to find $c \in (a, b)$ such that $g'(c) = 0$. This is equivalent to $g'(c) = f'(c) - d = 0$ or $f'(c) = d$. Since g is continuous on the interval, if g has a maximum or minimum point at c in (a, b) then $g'(c) = 0$. Assume that the maximum and minimum values of g are assumed at a and at b, the endpoints. Since $g'(a) < 0$, there must be a point, x, close to a such that $g(a) > g(x)$. Thus a cannot be the minimum point. Since $g'(b) > 0$ there must be a point, y, close to b such that $g(y) < g(b)$. Thus b cannot be the minimum point. This contradicts our assumption. Thus g must have a minimum point, c, in (a, b) and $g'(c) = 0$. □

8.1 Exercises

Exercise 8.1. Let $f(x) = \frac{1}{(x-1)^3}$ on $(1, \infty)$. Using the definition of the derivative, find $f'(c)$ for $c > 1$.

Exercise 8.2. Show that $\lim_{x\to 0} \cos\left(\frac{1}{x}\right)$ does not exist.

Exercise 8.3. Prove that $\lim_{x\to 0} x \sin\left(\frac{1}{x}\right) = 0$.

Exercise 8.4. Given that $\lim_{x\to 0} \frac{\sin x}{x} = 1$ and $\lim_{x\to 0} \frac{1-\cos x}{x} = 0$, compute the derivative of $\sin x$ using the definition.

Exercise 8.5. Using the definition of the derivative, show that $f(x) = \sqrt[3]{x}$ is not differentiable at $x = 0$. Find the equation of the tangent line to the graph of $f(x)$ at $(0, 0)$ (if possible).

Exercise 8.6. For a given function, $f(x)$, define $F(x) = \lim_{h\to 0} \frac{f(x+h)-f(x-h)}{2h}$.

a) Compute $F(x)$ for $f(x) = x^2$.

b) Now let $f(x) = |x|$ and compute $F(0)$.

c) Is $F(x)$ the derivative of $f(x)$? Why or why not?

Exercise 8.7. Let $f(x) = \begin{cases} x\sin\left(\frac{1}{x}\right) & \text{if } x \neq 0 \\ 0 & \text{if } x = 0 \end{cases}$. Show that f is continuous for all x. Show that $f'(0)$ does not exist.

Exercise 8.8. Let $f(x) = ax^2 + bx + c$ where a, b, c are real numbers. Using the limit definition of the derivative find $f'(x)$.

Exercise 8.9. Let $f(x) = \sqrt{2x}$. Using the limit definition of the derivative find $f'(x)$. What is the domain of $f'(x)$?

Exercise 8.10. Verify the Mean Value Theorem (Theorem 8.16) for $f(x) = x^3 - 6x + 2$ over $[-2, 0]$.

Exercise 8.11. Find the largest intervals on which $f(x) = x^3 - 3x$ is increasing.

Exercise 8.12. Find the local maximum and minimum points of $f(x) = |4 - x^2|$.

Exercise 8.13. For each example, precisely explain why the function you found fits the requested description *or* why no such function can exist. For each that does exist, provide a graph *and* a formula.

a) Give an example of a function $f : \mathbb{R} \to \mathbb{R}$ that is continuous everwhere but which is not differentiable at exactly 1 point.

b) Let $n \in \mathbb{N}$. Give an example of a function $f : \mathbb{R} \to \mathbb{R}$ that is continuous everywhere but which is not differentiable at exactly n points.

c) Give an example of a function $f : \mathbb{R} \to \mathbb{R}$ that is continuous everywhere but which is not differentiable at infinitely many points and which is differentiable at infinitely many (*other*) points.

d) Give an example of a function $f : \mathbb{R} \to \mathbb{R}$ that is differentiable everywhere but which is not continuous at exactly 1 point.

e) Let $n \in \mathbb{N}$. Give an example of a function $f : \mathbb{R} \to \mathbb{R}$ that is differentiable everywhere but which is not continuous at exactly n points.

f) Give an example of a function $f : \mathbb{R} \to \mathbb{R}$ that is differentiable everywhere but which is not continuous at infinitely many points and which is continuous at infinitely many (*other*) points.

g) Give an example of a function $f : (0, 1) \to \mathbb{R}$ that is continuous on its domain and that does not achieve an absolute maximum on $[0, 1]$.

h) Give an example of a function $f : [0, 1] \to \mathbb{R}$ that is continuous on its domain and that does not achieve an absolute maximum on $(0, 1)$.

Exercise 8.14. Verify the Generalized Mean Value Theorem (Theorem 8.18) for $f(x) = \frac{1}{x}$ and $g(x) = \frac{1}{x^2}$ over $[1, 4]$.

Exercise 8.15. Prove the Generalized Mean Value Theorem (Theorem 8.18) (Hint: Apply the Mean Value Theorem (Theorem 8.16) to $h(x) = [f(b) - f(a)]g(x) - [g(b) - g(a)]f(x)$. Make sure to show that all the hypotheses hold.)

Exercise 8.16. Use the Chain Rule, the Product Rule, and the fact that the derivative of $f(x) = \frac{1}{x}$ is $f'(x) = \frac{-1}{x^2}$ to prove the Quotient Rule, namely that $\left(\frac{f}{g}\right)' = \frac{f'(x)g(x) - f(x)g'(x)}{(g(x))^2}$ wherever f and g are differentiable and g is not 0.

Exercise 8.17. There are three hypotheses for Rolle's Theorem (Theorem 8.15), namely for $f : [a, b] \to \mathbb{R}$ (the real numbers):

a) f is continuous on $[a, b]$,

b) f is differentiable on (a, b), and

c) $f(a) = f(b)$.

Find examples of three different functions and intervals $[a, b]$ for which Rolle's Theorem fails. These functions should satisfy two of the hypotheses and fail for the third. Each function should fail a different hypothesis.

Chapter 9

The Riemann Integral

In integral calculus students are usually taught is the Riemann integral. The usual introduction to this integral is via the area problem — that is, How does one find the area of a region defined over a closed interval by a curve in the xy-plane? In this section we define the Riemann integral by defining underestimates (lower sums) and overestimates (upper sums) of such areas. Later we will define other estimates of these areas using Riemann sums. We close this chapter with the Fundamental Theorem of Calculus which brings together most of the material of this book.

Definition 9.1 (Partition of an Interval)**.** Let $[a, b]$ be a non-empty closed interval. A *partition of* $[a, b]$ is a set $\mathcal{P} = \{x_0, x_1, x_2, \ldots, x_n\}$ satisfying

$$a = x_0 < x_1 < \cdots < x_{n-1} < x_n = b.$$

The *kth subinterval of* $\mathcal{P}$ is the closed interval $[x_{k-1}, x_k]$ whose width is given by $\Delta x_k = x_k - x_{k-1}$. The *mesh* of $\mathcal{P}$ is the maximum of the subinterval lengths.

Note that $\mathcal{P}$ contains $n + 1$ distinct real numbers and that $x_0 = a$ and $x_n = b$. The partition $\mathcal{P}$ defines n closed subintervals.

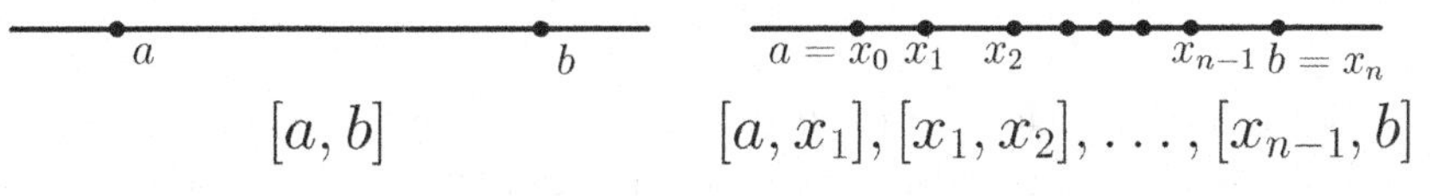

Figure 9.1

Remark 9.2. Figure 9.1 depicts a closed interval $[a, b]$ and a partition of $[a, b]$, given by $\{a = x_0, x_1, x_2, \ldots, x_{n-1}, x_n = b\}$. This partition defines n closed subintervals $[a, x_1], [x_1, x_2], \ldots, [x_{n-1}, b]$ with a union equal to the entire interval $[a, b]$. Adjacent subintervals intersect only in the common endpoint.

Example 9.3. Let $[a, b] = [3, 7]$ and let $\mathcal{P} = \{2, 4.5, 4.9, 5.6, 6, 7\}$. Then

$$\begin{aligned}
\Delta x_1 &= 4.5 - 3 = 1.5 \\
\Delta x_2 &= 4.9 - 4.5 = 0.4 \\
\Delta x_3 &= 5.6 - 4.9 = 0.7 \\
\Delta x_4 &= 6 - 5.6 = 0.4 \\
\Delta x_5 &= 7 - 6 = 1.0
\end{aligned}$$

The mesh of $\mathcal{P}$ is 1.5. Note that $73 = 1.5+0.4+0.7+0.4+1.0 = 4$. This means that the subintervals of the partition $\mathcal{P}$ cover the entire interval $[a, b]$, leaving no point out.

In general:

$$\sum_{k=1}^{n} \Delta x_k = x_n - x_0 = b - a.$$

Definition 9.4 (Lower and Upper Sums)**.** Let f be a bounded, real-valued function on $[a, b]$ and let P be a partition of the interval. For each $k = 1, 2, \ldots, n$ define

$$m_k = \text{glb}\{f(x) \mid x \in [x_{k-1}, x_k]\} \quad \text{and} \quad M_k = \text{lub}\{f(x) \mid x \in [x_{k-1}, x_k]\}.$$

Given these quantities, define the *lower and upper sums of f with respect to the partition $\mathcal{P}$* by:

$$L(f, P) = \sum_{k=1}^{m} m_k \Delta x_k \quad \text{(lower sum) and} \quad U(f, P) = \sum_{k=1}^{n} M_k \Delta x_k \quad \text{(upper sum)}.$$

Remark 9.5. It is not immediately obvious that the quantities m_k and M_k exist. That will be part of the work of the first theorem of the chapter. In the theorem, their existence follows from the requirement that the function, f, be bounded on the interval.

Theorem 9.6. Let f be a bounded function on $[a, b]$ and let $\mathcal{P} = \{x_0, x_1, x_2, \ldots, x_n\}$ be a partition of the interval. The quantities m_k and M_k are defined and $L(f, P) \leq U(f, P)$.

Proof. Since f is a bounded function on each subinterval the greatest lower bound and least upper bound of the set of its values on the subintervals both exist by the Completeness Axiom (Axiom 6). The set $\{f(x)|x \in [x_{k-1}, x_k]\}$ is non-empty since it contains the number $f(x_k)$ and is bounded both above and below since f is a bounded function. Since $m_k \leq M_k$ for each k, it follows that

$$L(f, \mathcal{P}) = \sum_{k=1}^{n} m_k \Delta x_k \leq \sum_{k=1}^{n} M_k \Delta x_k = U(f, \mathcal{P}). \qquad \square$$

Example 9.7. Let $[a, b] = [3, 7]$ and $\mathcal{P} = \{3, 4.5, 4.9, 5.6, 6, 7\}$ as in Example 9.3. Finally let $f(x) = x^2 + 2x$ on $[3, 7]$. Then we can calculate the upper and lower bounds of $f(x)$ over each subinterval in P, and create a table to track our calculations. This makes the calculation of $L(f, \mathcal{P})$

and $U(f, \mathcal{P})$ tidy.

k	$[x_{k-1}, x_k]$	Δx_k	m_k	$m_k \Delta x_k$	M_k	$M_k \Delta x_k$
1	$[3, 4.5]$	1.5	15	22.5	29.25	43.875
2	$[4.5, 4.9]$	0.4	29.25	11.7	33.81	13.524
3	$[4.9, 5.6]$	0.7	33.81	23.667	42.56	29.792
4	$[5.6, 6]$	0.4	42.56	17.024	48	19.2
5	$[6, 7]$	1	48	48	63	63
			$L(f, \mathcal{P}) = 168.62$		$U(f, \mathcal{P}) = 216.62$	

The upper and lower sums come from adding entries in the pertinent columns. Here, $L(f, \mathcal{P}) = 168.62$ and $U(f, \mathcal{P}) = 216.62$.

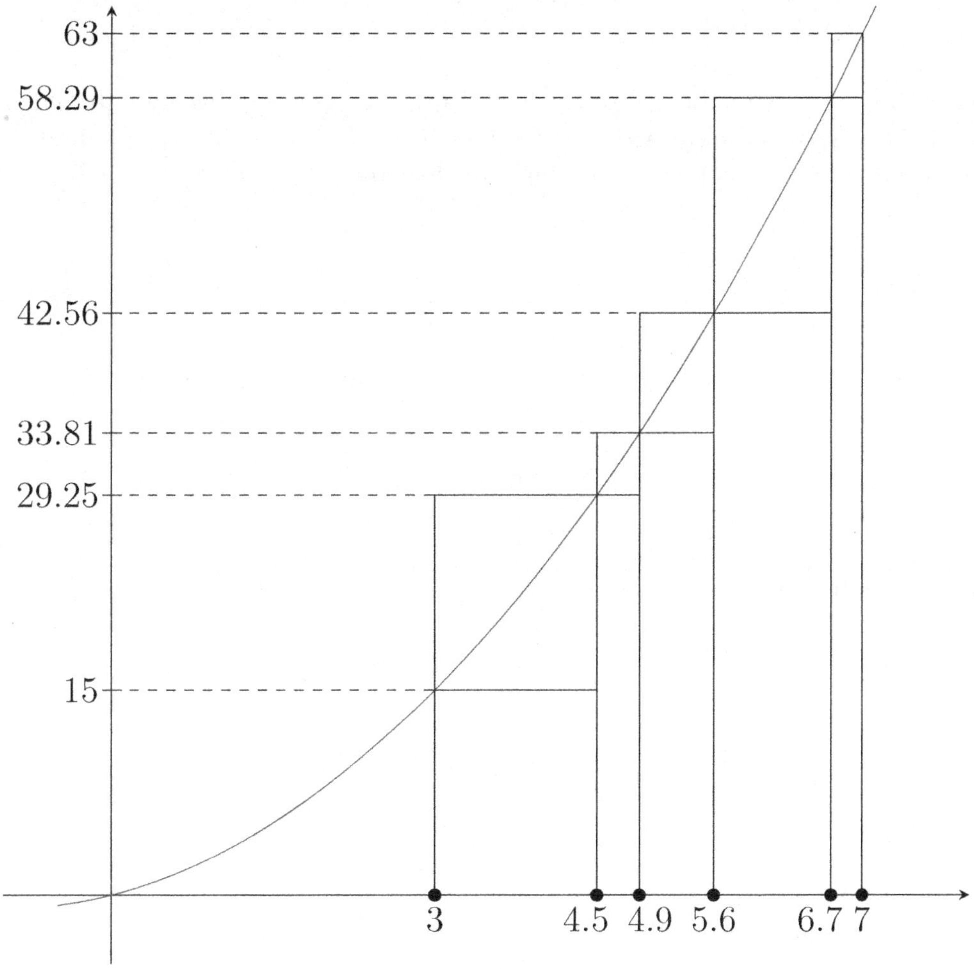

Figure 9.2

Definition 9.8 (Refinement of a Partition). Let $\mathcal{P}$ and $\mathcal{Q}$ be partitions of $[a, b]$. We call $\mathcal{Q}$ a *refinement of* $\mathcal{P}$ if $\mathcal{P} \subset \mathcal{Q}$.

It initially sounds funny that the larger set is the refinement, but think about the picture of $\mathcal{P}$ and $\mathcal{Q}$ on the real line. The subintervals of $\mathcal{Q}$ are *smaller* than those of $\mathcal{P}$, because there are more endpoints in the set $\mathcal{Q}$.

Theorem 9.9. Let f be a bounded function on $[a, b]$ and let $\mathcal{P}$ and $\mathcal{Q}$ be partitions of $[a, b]$ with $\mathcal{Q}$ a refinement of $\mathcal{P}$. Then $L(f, \mathcal{P}) \leq L(f, \mathcal{Q}) \leq U(f, \mathcal{Q}) \leq U(f, \mathcal{P})$.

Proof. We will prove the leftmost inequality. The middle inequality has already been shown and the rightmost is very similar to the leftmost.

Since both $\mathcal{P}$ and $\mathcal{Q}$ are finite sets it suffices to prove the inequality for the case in which $\mathcal{Q}$ contains one more point than $\mathcal{P}$. Let $\mathcal{Q} = \mathcal{P} \cup \{y\}$ where $x_{k-1} < y < x_k$. The lower sums over $\mathcal{P}$ and $\mathcal{Q}$ are equal except over the subinterval $[x_{k-1}, x_k]$.

$$m_k = \min\{\text{glb}\{f(x)|x \in [x_{k-1}, y]\}, \text{glb}\{f(x)|x \in [y, x_k]\}\}$$

$$\begin{aligned} m_k \Delta x_k &= m_k \left((y - x_{k-1}) + (x_k - y)\right) \\ &\leq \text{glb}\left\{f(x)|x \in [x_{k-1}, y]\right\}(y - x_{k-1}) + \text{glb}\left\{f(x)|x \in [y, x_k]\right\}(x_k - y) \end{aligned}$$

The right hand side is the contribution to the lower sum over $\mathcal{Q}$ of the two subintervals $[x_{k-1}, y]$ and $[y, x_k]$. The left hand side is the contribution to the lower sum over $\mathcal{P}$ from the subinterval $[x_{k-1}, y]$. Since all the other terms in the two lower sums are the same it follows that $L(f, \mathcal{P}) \leq L(f, \mathcal{Q})$. □

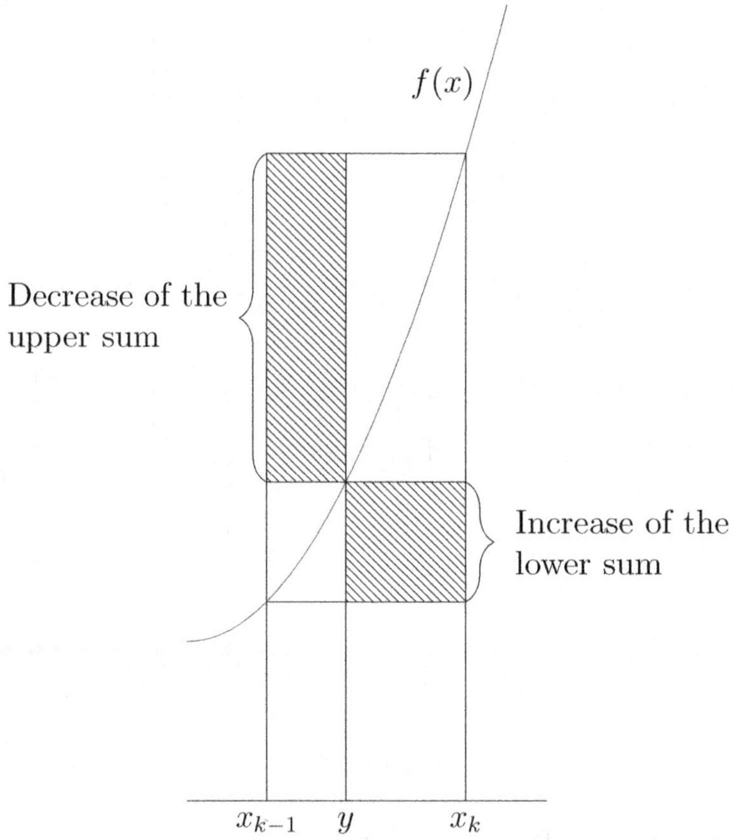

Figure 9.3

Remark 9.10. Notice that the greatest lower bound of the value of f is greater on the interval $[y, x_k]$ than it is on the interval $[x_{k-1}, x_k]$. This leads to a greater lower sum over the new partition, the increase being the area of the shaded portion over the second interval. A similar analysis holds for the decrease in the upper sum by the area in the other shaded rectangle.

Theorem 9.11. Let $\mathcal{P}$ and $\mathcal{Q}$ be partitions of $[a, b]$. Then $\mathcal{P} \cup \mathcal{Q}$ is a refinement of both $\mathcal{P}$ and $\mathcal{Q}$. Further, $L(f, \mathcal{P}) \leq U(f, \mathcal{Q})$.

Proof. Let $\mathcal{R} = \mathcal{P} \cup \mathcal{Q}$. By the previous theorem $L(f, \mathcal{P}) \leq L(f, \mathcal{R})$ and also $U(f, \mathcal{R}) \leq U(f, \mathcal{Q})$ so the result is a matter of putting the two inequalities together to get

$$L(f, \mathcal{P}) \leq L(f, \mathcal{R}) \leq U(f, \mathcal{R}) \leq U(f, \mathbb{Q}). \qquad \square$$

Remark 9.12. At first glance, Theorems 9.9 and 9.11 seem to say the same thing. In Theorem 9.9, the upper and lower sums with respect to a partition $\mathcal{P}$ and *a refinement* are compared, but in Theorem 9.11, the upper and lower sums of two *potentially incomparable* partitions are compared. Even if the partitions themselves are not comparable, a lower sum is still less than or equal to an upper sum.

Definition 9.13 (The Riemann Integral)**.** Let f be a bounded function on $[a, b]$ and let $\mathscr{P}$ be the set of all partitions of the interval. Define the *lower integral of f over* $[a, b]$ by $L(f) = \text{lub}\{L(f, \mathcal{P}) | \mathcal{P} \in \mathscr{P}\}$ and the *upper integral of f over* $[a, b]$ by $U(f) = \text{glb}\{U(f, \mathcal{P}) | \mathcal{P} \in \mathscr{P}\}$.

We say that f *is Riemann-integrable over* $[a, b]$ if $L(f) = U(f)$ and denote the common value by $\int f$ or $\int\limits_a^b f(x)\, dx$.

Remark 9.14. Again it is not obvious that the upper and lower integrals exist. Again their existence comes from the boundedness of f. When the function f is unbounded we enter the realm of what is known as the Improper Integral.

Theorem 9.15. Let f be a bounded function on $[a, b]$. Then both the upper and lower integrals of f exist.

Proof. The set of all lower sums of f and the set of all upper sums over all partitions of $[a, b]$ are both clearly non-empty. Let $\mathcal{P} = \{a, b\}$ be a partition of $[a, b]$. Then both $L(f, \mathcal{P})$ and $U(f, \mathcal{P})$ exist. Thus the sets of lower sums and upper sums are non-empty. By Theorem 9.11 above the set of lower sums is bounded above by any upper sum and the set of upper sums is bounded below by any lower sum. Thus by the Completeness Axiom (Axiom 6) the lower integral and the upper integral both exist. $\square$

The following theorem is extremely useful for determining if a function is Riemann-integrable over a given closed interval.

Theorem 9.16. A function f is Riemann-integrable over $[a, b]$ if and only if for each $\varepsilon > 0$ there is a partition, $\mathcal{P}_\varepsilon$, such that $U(f, \mathcal{P}_\varepsilon) - L(f, \mathcal{P}_\varepsilon) < \varepsilon$.

Proof. For the forward direction: We assume that f is Riemann integrable over $[a, b]$ and hence that $L(f) = U(f)$. Let $\varepsilon > 0$ be given. Recall Theorem 3.5 and its corollary, Corollary 3.6. Since $L(f) = \text{lub}\{L(f, \mathcal{P}) | \mathcal{P} \in \mathscr{P}\}$ there is a partition $\mathcal{P}_1$ satisfying $L(f) - \frac{\varepsilon}{2} < L(f, \mathcal{P}_1)$. Similarly since $U(f) = \text{glb}\{U(f, \mathcal{P}) | \mathcal{P} \in \mathscr{P}\}$ there is a partition $\mathcal{P}_2$ satisfying $U(f) + \frac{\varepsilon}{2} > U(f, \mathcal{P}_2)$. Let $\mathcal{P}_\varepsilon = \mathcal{P}_1 \cup \mathcal{P}_2$. Then combining inequalities and using the fact that $\mathcal{P}_\varepsilon$ is a refinement of $\mathcal{P}_1$ and $\mathcal{P}_2$ yields

$$L(f) - \frac{\varepsilon}{2} < L(f, \mathcal{P}_1) \leq U(f, \mathcal{P}_2) < U(f) + \frac{\varepsilon}{2}.$$

Then

$$U(f,\mathcal{P}_\varepsilon) - L(f,\mathcal{P}_\varepsilon) < U(f) + \frac{\varepsilon}{2} - \left(L(f) - \frac{\varepsilon}{2}\right) = \varepsilon$$

since $U(f) = L(f)$.

For the second direction: Let $\varepsilon > 0$ be given. Let $\mathcal{P}_\varepsilon$ be a partition satisfying $U(f,\mathcal{P}_\varepsilon) - L(f,\mathcal{P}_\varepsilon) < \varepsilon$. We have the following chain of inequalities:

$$L(f,\mathcal{P}_\varepsilon) \le L(f) \le U(f) \le U(f,\mathcal{P}_\varepsilon).$$

Since the outer two are within ε of each other so are the inner two. Thus we have $|U(f) - L(f)| < \varepsilon$ for all $\varepsilon > 0$. Thus they are equal and f is Riemann integrable. □

With this theorem in hand we can show that two large classes of bounded functions are Riemann integrable.

Theorem 9.17. If f is continuous on $[a,b]$ then it is Riemann-integrable over $[a,b]$.

Proof. If f is continuous on $[a,b]$, then it is uniformly continuous there. (See Theorem 7.24.) Let $\varepsilon > 0$ be given. Then by uniform continuity there is a $\delta > 0$ such that if $|x-y| < \delta$ and $x, y \in [a,b]$ then $|f(x) - f(y)| < \frac{\varepsilon}{b-a}$. Let $\mathcal{P}_\varepsilon$ be any partition of $[a,b]$ such that $\Delta x_k < \delta$ for all k. Since f takes on its maximum and minimum values at points in every closed subinterval (by the Extreme Value Theorem, Theorem 7.20) we have that

$$M_k - m_k = |f(u_k) - f(v_k)| < \frac{\varepsilon}{b-a}$$

where u_k and v_k are points in $[x_{k-1}, x_k]$ where f takes on its maximum and minimum values. Then

$$\begin{aligned} U(f,\mathcal{P}_\varepsilon) - L(f,\mathcal{P}_\varepsilon) &= \sum_{k=1}^{n} (M_k - m_k)\Delta x_k \\ &< \sum_{k=1}^{n} \left(\frac{\varepsilon}{b-a}\right) \Delta x_k \\ &= \left(\frac{\varepsilon}{b-a}\right) \sum_{k=1}^{n} \Delta x_k \\ &= \left(\frac{\varepsilon}{b-a}\right)(b-a) = \varepsilon. \end{aligned}$$

Hence f is Riemann-integrable on the interval. □

Example 9.18. Let $f(x) = 2x^3 + 3x$ on $[0,2]$ and let $\varepsilon = 0.01$. Use Theorem 9.17 to find a partition $\mathcal{P}$ on $[0,2]$ so that $|U(f,\mathcal{P}) - L(f,\mathcal{P})| < 0.01$.

First we need to find a $\delta > 0$ for the given ε that demonstrates the uniform continuity of f over the interval. If $x, y \in [0,2]$ then

$$\begin{aligned} |f(x) - f(y)| &= |2x^2 + 3x - 2y^2 - 3y| \\ &= |x-y||2x + 2y + 3| \\ &\le 11|x-y| \end{aligned}$$

since $|2x+2y+3| \leq 11$ for all x and y in the interval. Given $\varepsilon = 0.01$, if

$$\delta = \frac{\varepsilon}{22} = \frac{0.01}{22} = 0.00045454545\cdots$$

then whenever $|x-y| < \delta$, it follows that $|f(x) - f(y)| < \frac{\varepsilon}{2} = \frac{\varepsilon}{2-0}$. Choosing $\delta = 0.0002$ will guarantee that $|f(x) - f(y)| < \frac{0.01}{2}$.

Choose a partition of $[0, 2]$ such that each subinterval has length less than 0.0004. For example the uniform partition of $[0, 2]$ that breaks the interval into $10,000$ equal-length subintervals will result in a subinterval width of $\Delta x_k = 0.0002$ for all k, less than our needed δ.

Theorem 9.19. If f is monotonic on $[a, b]$, then f is Riemann-integrable on $[a, b]$.

Proof. Assume that f is non-decreasing on $[a, b]$. Let $\varepsilon > 0$ be given. Let n be a natural number satisfying $\frac{(b-1)(f(b)-f(a))}{n} < \varepsilon$. Let $\mathcal{P}_\varepsilon$ be any partition of $[a, b]$ with exactly n subintervals of equal length. Then $\Delta x_k = \frac{b-a}{n}$ for all k. Consequently,

$$\begin{aligned} U(f, \mathcal{P}_\varepsilon) - L(f, \mathcal{P}_\varepsilon) &= \sum_{k=1}^{n} (f(x_k) - f(x_{k-1})\Delta x_k \\ &= \frac{b-a}{n} \sum_{k=1}^{n} (f(x_k) - f(x_{k-1})) \\ &= \frac{(b-a)(f(b) - f(a))}{n} < \varepsilon. \end{aligned}$$

This proof depends on the fact that the maximum value of f occurs at the right endpoint of each subinterval and the minimum value at the left endpoint. Thus

$$\begin{aligned} \sum_{k=1}^{n} (f(x_k) - f(x_{k-1})) &= (f(x_1) - f(a)) + (f(x_2) - f(x_1)) + \cdots + (f(b) - f(x_{n-1})) \\ &= f(b) - f(a). \end{aligned}$$

□

Example 9.20. Repeat Example 9.18 except use Theorem 9.19 to determine the partition. The quantity $U(f, \mathcal{P}_\varepsilon) - L(f, \mathcal{P}_\varepsilon)$ equals

$$U(f, \mathcal{P}_\varepsilon) - L(f, \mathcal{P}_\varepsilon) = \frac{(b-a)(f(b) - f(a))}{n} = \frac{2 \cdot 14}{n} = \frac{28}{n}.$$

This must be less than 0.01. Then

$$\frac{28}{n} < 0.01 = \frac{1}{100},$$

or $2,800 < n$. Choose $n = 3,000$.

Theorem 9.21. Suppose that f is integrable on $[a, b]$ and that $a < c < b$. Then f is integrable on both $[a, c]$ and $[c, b]$ and further $\int_a^b f = \int_a^c f + \int_a^b f$.

Proof. The proof follows simply from including c in every partition $\mathcal{P}$ of $[a, b]$ and making the appropriate computations. That is, given a partition $\mathcal{P}$ let $\mathcal{Q} = \mathcal{P} \cup \{c\}$ be a refinement of $\mathcal{P}$. Let $\mathcal{Q}_1 = \mathcal{Q} \cap [a, b]$ and $\mathcal{Q}_2 = \mathcal{Q} \cap [c, b]$. Then $\mathcal{Q}_1$ and $\mathcal{Q}_2$ are partitions of $[a, c]$ and $[c, b]$ respectively and the equality of integrals follows from the fact that f is Riemann integrable on $[a, b]$. □

Theorem 9.22. Let f and g be functions integrable on $[a, b]$. Then

(1) $f + g$ is integrable on $[a, b]$ and $\int_a^b f + g = \int_a^b f + \int_a^b g$,

(2) if $r \in \mathbb{R}$ then $\int_a^b rf = r\int_a^b f$,

(3) if $m \leq f \leq M$ (for constants $m, M \in \mathbb{R}$) on $[a, b]$ then $m(b - a) \leq \int_a^b \leq M(b - a)$,

(4) if $f \leq g$ on $[a, b]$ then $\int_a^b f \leq \int_a^b g$, and

(5) $|f|$ is integrable with $\left|\int_a^b f\right| \leq \int_a^b |f|$.

Proof. (1) The proof follows from the following facts on each subinterval of any partition:

$$\text{glb}\{f(x)|x \in [x_{k-1}, x_k]\} + \text{glb}\{g(x)|x \in [x_{k-1}, x_k]\} \leq \text{glb}\{f(x) + g(x)|x \in [x_{k-1}, x_k\}$$

and

$$\text{lub}\{f(x) + g(x)|x \in [x_{k-1}, x_k\} \leq \text{lub}\{f(x)|x \in [x_{k-1}, x_k]\} + \text{lub}\{g(x)|x \in [x_{k-1}, x_k]\}.$$

These inequalities imply that

$$L(f, \mathcal{P}) + L(g, \mathcal{P}) \leq L(f + g, \mathcal{P}) \leq U(f + g, \mathcal{P}) \leq U(f, \mathcal{P}) + U(g, \mathcal{P})$$

for all partitions $\mathcal{P}$ of $[a, b]$. For every $\varepsilon > 0$ there exists a partition $\mathcal{P}_1$ of $[a, b]$ such that

$$U(f, \mathcal{P}_1) - L(f, \mathcal{P}_1) < \frac{\varepsilon}{2}$$

and similarly there is a partition $\mathcal{P}_2$ such that

$$U(g, \mathcal{P}_1) - L(g, \mathcal{P}_1) < \frac{\varepsilon}{2},$$

by Theorem 9.16. Letting $\mathcal{P} = \mathcal{P}_1 \cup \mathcal{P}_2$, we now have a partition of $[a, b]$ such that

$$U(f + g, \mathcal{P}) - L(f + g, \mathcal{P}) < \varepsilon$$

and then Theorem 9.16 allows us to conclude that $f + g$ is integrable.

(2) We need to consider three cases here. The simplest is the case $r = 0$. Then both sides are equal to 0. Now suppose that $r > 0$. Let $\mathcal{P}$ be any partition of the given interval with m_k and M_k, the bounds on $f(x)$ on $[x_{k-1}, x_x]$, determined for that partition. Then

$$rm_k = \text{glb}\{rf(x)|x \in [x_{k-1}, x_k]\} \quad \text{and} \quad rM_k = \text{lub}\{rf(x)|x \in [x_{k-1}, x_k]\}.$$

Then $L(rf, \mathcal{P}) = rL(f, \mathcal{P})$ and $U(rf, \mathcal{P}) = rU(f, \mathcal{P})$. Then given $\varepsilon > 0$, if $\mathcal{P}$ is a partition such that

$$|U(f, \mathcal{P}) - L(f, \mathcal{P}) < \frac{\varepsilon}{r}$$

then

$$|U(rf, \mathcal{P}) - L(rf, \mathcal{P})| < \varepsilon.$$

If $r < 0$ the roles of m_k and M_k are reversed and the proof follows.

(3) Let $\mathcal{P}$ be any partition of $[a, b]$. Consider the constant *functions* m and M on the interval. Then

$$L(m, \mathcal{P}) = U(m, \mathcal{P}) = \int_a^b m = m(b - a)$$

and

$$L(M, \mathcal{P}) = U(M, \mathcal{P}) = \int_a^b M = M(b - a).$$

Since $m \le f \le M$ it follows that

$$L(m, \mathcal{P}) \le L(f, \mathcal{P}) \le L(M, \mathcal{P})$$

and

$$U(m, \mathcal{P}) \le U(f, \mathcal{P}) \le U(M, \mathcal{P}).$$

Thus

$$m(b - a) \le L(f, \mathcal{P}) \le \int_a^b f \le U(f, \mathcal{P}) \le M(b - a).$$

(4) Since $f \le g$ it follows that $L(f, \mathcal{P}) \le L(g, \mathcal{P})$ and $U(f, \mathcal{P}) \le U(g, \mathcal{P})$ for any partition P. This forces $\int_a^b f \le \int_a^b g$.

(5) Let $\varepsilon > 0$ be given and let $\mathcal{P}$ be a partition such that $U(f, \mathcal{P}) - L(f, \mathcal{P}) \le \varepsilon$. This holds because f is assumed integrable. Let m_k and M_k be the numbers for f and $\mathcal{P}$, that is, the greatest lower bounds and least upper bounds over the various subintervals. Then

$$U(f, \mathcal{P}) - L(f, \mathcal{P}) = \sum_{k=1}^{n} (M_k - m_k)\Delta x_k.$$

We now consider what happens with the function $|f|$. We will show that the analogous difference for $|f|$ is less than or equal to that for f. This will require three separate cases.

- **Case 1:** $0 \le m_k \le M_k$**.** The numbers on each subinterval for $|f|$ are the same as for f (since every $f(x)$ is greater than or equal to m_k and thus greater than or equal to 0), thus $(M_k - m_k)\Delta x_k$ is unchanged for $|f|$.

- **Case 2:** $m_k \le M_k \le 0$**.** Now M_k and $-m_k$ are the greatest lower bound and least upper bound of the function values of $|f|$ over the subinterval in question. Notice that

$$-m_k - (-M_k) = M_k - m_k.$$

  Again the appropriate terms in the difference between the upper and lower sums are unchanged.

- **Case 3:** $m_k \le 0 \le M_k$**.** Let $Z_k = \max\{m_k, M_k\}$. The number Z_k is the least upper bound of the values of $|f|$ over the kth subinterval. Let z_k be the greatest lower bound of the values of $|f|$ over the same subinterval. All we know is that $0 \le z_k \le Z_k$. Then the difference of the least upper bound and greatest lower bound over the subinterval for $|f|$ is $Z_k - z_k$. It follows from what we know that

$$Z_k - z_k \le Z_k \le M_k + |m_k| = M_k - m_k$$

  since the greatest lower bound of the function values is less than or equal to 0.

Thus the appropriate term in $U(|f|,\mathcal{P})-L(|f|,\mathcal{P})$ is less than the corresponding term in $U(f,\mathcal{P})-L(f,\mathcal{P})$ for each k and consequently, we can make $U(|f|,\mathcal{P})-L(|f|,\mathcal{P})$ arbitrarily small. This shows that $|f|$ is Riemann integrable.

Since we know that for all x in $[a,b]$,

$$-|f(x)| \le f(x) \le |f(x)|,$$

it follows by (4)) that

$$\int_a^b -|f| \le -\int_a^b |f| \le \int_a^b f \le \int_a^b |f|.$$

Apply the absolute value to the integrals and we have $\left|\int_a^b f\right| \le \int_a^b |f|$. □

Theorem 9.23. Suppose that f is integrable on $[a,c]$ for every c satisfying $a<c<b$ and bounded on $[a,b]$. Then f is integrable on $[a,b]$.

Proof. Let $\varepsilon>0$ be given and let M satisfy $|f(x)|\le M$ for all $x\in[a,b]$. Choose c satisfying $a<c<b$ such that $b-c<\frac{\varepsilon}{4M}$. Let $\mathcal{P}_1$ be a partition of $[a,c]$ such that $U(f,\mathcal{P}_1)-L(f,\mathcal{P}_1)<\frac{\varepsilon}{2}$ which is possible by the hypotheses of the theorem. Let $\mathcal{P}_\varepsilon=\mathcal{P}_1\cup\{b\}$. Then

$$U(f,\mathcal{P}_\varepsilon)-L(f,\mathcal{P}_\varepsilon)\le U(f,\mathcal{P}_1)=L(f,\mathcal{P}_1)+2M\left(\frac{\varepsilon}{4M}\right)<\frac{\varepsilon}{2}+\frac{\varepsilon}{2}=\varepsilon.$$

This is true because the difference between the least upper bound and greatest lower bound of the values of f on the rightmost subinterval is less than or equal to $2M$. □

Up to this point we have used the integral defined by lower sums and upper sums. There is an equivalent formulation in terms of Riemann sums, sums in which the terms are the products of function values and lengths of intervals. The precise definition is as follows.

Definition 9.24 (Riemann Sum)**.** Let f be bounded on $[a,b]$ and let $\mathcal{P}$ be a partition of $[a,b]$. A *marking of* $\mathcal{P}$ is a collection of points taken from the subintervals of $\mathcal{P}$, one for each subinterval. Let $C=\{c_1,c_2,\dots,c_n\}$ where $c_k\in[x_{k-1},x_k]$. The set C is a marking of $\mathcal{P}$. The *Riemann Sum corresponding to* f, $\mathcal{P}$ *and* C is given by $R(\mathcal{P},C,f)=\sum_{k=1}^{n} f(c_k)\Delta x_k$.

Example 9.25. Let $f(x)=x^2+x$ on $[1,2]$ and let $\mathcal{P}=\{1,1.2,1.7,2\}$ be a partition of $[1,2]$. Let $C=\{1.1,1.4,1\}$ be a marking of $\mathcal{P}$. Then

$$\begin{aligned}
R(\mathcal{P},C,f) &= \sum_{k=1}^{n} f(c_k)\Delta x_k\\
&= \left((1.1)^2+1.1\right)(0.2)+\left((1.4)^2+1.4\right)(0.5)+\left((1.1)^2+1\right)(0.3)\\
&= 2.742
\end{aligned}$$

In other words, the Riemann Sum corresponding to $\mathcal{P}$, C, and f equals 2.742.

Theorem 9.26. Let f be Riemann integrable on $[a,b]$. Let $\varepsilon>0$ be given and let $\mathcal{P}_\varepsilon$ be a partition of $[a,b]$ such that $U(f,\mathcal{P}_\varepsilon)-L(f,\mathcal{P}_\varepsilon)<\varepsilon$. Then for every marking C of $\mathcal{P}_\varepsilon$, $\left|\int_a^b f - R(\mathcal{P}_\varepsilon,C,f)\right|<\varepsilon$. In other words: The Riemann sums closely approximate the integral.

Proof. Since $m_k \leq f(c_k) \leq M_k$ on each subinterval of $\mathcal{P}_\varepsilon$, it follows that

$$L(f, \mathcal{P}_\varepsilon) \leq R(\mathcal{P}_\varepsilon, C, f) \leq U(f, \mathcal{P}_\varepsilon).$$

We know from earlier work that $L(f, \mathcal{P}_\varepsilon) \leq \int_a^b f \leq U(f, \mathcal{P}_\varepsilon)$. Thus

$$\left| \int_a^b f - R(\mathcal{P}_\varepsilon, C, f) \right| < U(f, \mathcal{P}_\varepsilon) - L(f, \mathcal{P}_\varepsilon) < \varepsilon. \qquad \square$$

We now come to the Fundamental Theorem of Calculus whose proof uses almost everything we have developed. The first part of the theorem is essential for computing integrals in calculus while part 2 is at the heart of the subject of differential equations. The theorem explores the relations between integration and differentiation. It shows that these are inverse processes.

Theorem 9.27 (The Fundamental Theorem of Calculus)**.**

(1) Suppose that f is integrable on $[a, b]$ and that F is a function on $[a, b]$ that satisfies $F'(x) = f(x)$ for all x in $[a, b]$. Then $\int_a^b f = F(b) - F(a)$.

(2) Suppose that f is integrable on $[a, b]$ and that F is defined by $F(x) = \int_a^x f$ for all x in $[a, b]$. Then F is continuous on $[a, b]$ and if f is continuous at c satisfying $a < c < b$ then F is differentiable at c and $F'(c) = f(c)$.

Proof. For (1): Let $\varepsilon > 0$ be given and let $\mathcal{P}_\varepsilon$ be a partition of $[a, b]$ such that $U(f, \mathcal{P}_\varepsilon) - L(f, \mathcal{P}_\varepsilon) < \varepsilon$. We apply the Mean Value Theorem (Theorem 8.16) to the function $F(x)$ over each subinterval $[x_{k-1}, x_k]$ of $\mathcal{P}_\varepsilon$. (Why does the Mean Value Theorem apply on each of these intervals?) Thus there is a $c_k \in (x_{k-1}, x_k)$ such that

$$f(c_k) = F'(c_k) = \frac{F(x_k) - F(x_{k-1})}{x_k - x_{k-1}}.$$

We rewrite this as $F(x_k) - F(x_{k-1}) = f(c_k)\Delta x_k$. The set $C = \{c_1, c_2, \ldots, c_n\}$ is a marking of $\mathcal{P}_\varepsilon$ and the Riemann sum

$$R(\mathcal{P}_\varepsilon, C, f) = \sum_{k=1}^{n} f(c_k)\Delta x_k = \sum_{k=1}^{n} (F(x_k) - F(x_{k-1})) = F(b) - F(a)$$

since the next-to-the-last term is a telescoping sum. Thus for every $\varepsilon > 0$ there is a marking C such that $\left| R(\mathcal{P}_\varepsilon, C, f) - \int_a^b f \right| < \varepsilon$ and hence that $\left| F(b) - F(a) - \int_a^b f \right| < \varepsilon$ for every $\varepsilon > 0$. Since this is true for all positive ε it follows that $\int_a^b f = F(b) - F(a)$.

For (2): This part is harder to prove. It also has two separate parts. First we prove the continuity of the function $F(x)$. Let $c \in [a, b]$. For any $x \in [a, b]$,

$$|F(x) - F(c) = \left| \int_c^x f \right| \leq M|x - c|$$

where M is a bound on f, that is $|f(x)| \leq M$ on $[a, b]$. Let $\varepsilon > 0$ be given and choose $\delta = \frac{\varepsilon}{M}$. Then if $|x - c| < \delta = \frac{\varepsilon}{M}$ we have $|F(x) - F(c)| \leq M|x - c| < \varepsilon$. Thus F is continuous at each c.

Now we assume that f is continuous at c and show that $F'(c) = f(c)$. Consider the following quantity:

$$\left|\frac{F(x)-F(c)}{x-c}-f(c)\right|.$$

If the limit of this quantity as x approaches c is 0 then we are done.

$$\begin{aligned}\left|\frac{F(x)-F(c)}{x-c}-f(c)\right| &= \left|\frac{1}{x-c}\left(\int_c^x f\right)-f(c)\right| \\ &= \left|\frac{1}{x-c}\left(\int_c^x f(t)\ \mathrm{d}t\right)-\frac{1}{x-c}\left(\int_c^x f(c)\ \mathrm{d}t\right)\right| \\ &= \left|\frac{1}{x-c}\right|\left|\int_c^x f(t)-f(c)\ \mathrm{d}t\right|\end{aligned}$$

Since we have assumed that f is continuous at c we know that for a given $\varepsilon > 0$ there is a $\delta > 0$ such that if $|x-c| < \delta$ then $|f(x)-f(c)| < \varepsilon$.

Going back to our original quantity and assuming that $|x-c| < \delta$ we can conclude that

$$\begin{aligned}\left|\frac{F(x)-F(c)}{x-c}-f(c)\right| &= \left|\frac{1}{x-c}\right|\left|\int_c^x f(t)-f(c)\ \mathrm{d}t\right| \\ &\leq \left|\frac{1}{x-c}\right|(\varepsilon\,|x-c|)=\varepsilon.\end{aligned}$$

We have used parts (3) and (5) of Theorem 9.22. Thus we have shown that

$$\lim_{x\to c}\left|\frac{F(x)-F(c)}{x-c}-f(c)\right|=0$$

or equivalently $F'(c)=f(c)$. □

Theorem 9.28 (The Mean Value Theorem for Integrals)**.** Let f be continuous on $[a,b]$. Then there is a c satisfying $a < c < b$ such that $f(c)(b-a)=\int_a^b f$.

9.1 Exercises

Exercise 9.1. Let $f : \mathbb{R} \to \mathbb{R}$ be defined by $f(x) = x^2$ if x is rational and $f(x) = x^3$ if x is irrational. Find each of the following.

a) $\text{lub}\{f(x)|x \in [0, 0.5]\}$

b) $\text{glb}\{f(x)|x \in [0, 0.5]\}$

c) $\text{lub}\{f(x)|x \in [0.5, 1.5]\}$

d) $\text{glb}\{f(x)|x \in [0.5, 1.5]\}$

e) $\text{lub}\left\{f(x)|x \in \left[\frac{\sqrt{2}}{3}, \frac{\sqrt{2}}{2}\right]\right\}$

f) $\text{glb}\left\{f(x)|x \in \left[\frac{\sqrt{2}}{3}, \frac{\sqrt{2}}{2}\right]\right\}$

Exercise 9.2. Let $f(x) = x^2$ and $P = \{0, 0.2, 0.5, 0.7, 1\}$. Compute $L(f, P)$ and $U(f, P)$. Then let $P_1 = \{0, 0.2, 0.5, 0.6, 0.7, 0.9, 1\}$ be a refinement of P. Compute $L(f, P_1)$ and $U(f, P_1)$. Compare the two lower sums and the two upper sums.

Exercise 9.3. Let $f(x) = x^2$ and $P_n = \left\{0, \frac{1}{n}, \frac{2}{n}, \frac{3}{n}, \ldots, 1\right\}$. Compute $L(f, P_n)$ and $U(f, P_n)$.

Exercise 9.4. Let $P = \left\{0, \frac{1}{4}, \frac{1}{2}, \frac{3}{4}, 1\right\}$ and let f be defined by:

$$f(x) = \begin{cases} 0 & \text{if } 0 \le x < \frac{1}{4} \\ 1 & \text{if } \frac{1}{4} \le x \le \frac{1}{2} \\ 2 & \text{if } \frac{1}{2} < x < \frac{3}{4} \\ 3 & \text{if } \frac{3}{4} \le x \le 1 \end{cases}.$$

Find m_1, m_2, m_3, m_4, M_1, M_2, M_3, M_4, $L(f, P)$, and $U(f, P)$.

Exercise 9.5. Prove: If $f(x)$ is Riemann integrable on $[a, b]$ and r is a real number, then $rf(x)$ is also Riemann integrable on $[a, b]$ and $\int_a^b rf(x)\, dx = r \int_a^b f(x)\, dx$.

Exercise 9.6. Let $f(x) = 0$ if x is rational and $f(x) = x$ if x is irrational with f defined on $[0, 1]$. What are $L(f)$ and $U(f)$?

Exercise 9.7. Prove that if f is Riemann integrable on $[a, b]$ and $m \le f(x) \le M$ for all x in the interval then $m(b - a) \le \int_a^b f \le M(b - a)$.

Exercise 9.8. Let n be a natural number and let $f(x) = x^3$ on $[0, 2]$ except that $f(x) = 0$ at all points of the form k/n where $k, n \in \mathbb{N}$ and $0 \le k \le 2n$; i.e., $f(x) = 0$ at all rational numbers with a denominator of n in $[0, 2]$. Is f Riemann integrable on $[0, 2]$? If so, find the integral of f over the interval.

Exercise 9.9. Let $f(x)$ be a continuous, increasing function on $[a, b]$ with $f(a) = 0$ and $f(b) = M$. Let $P_n = \left\{a, a + \frac{b-a}{n}, a + \frac{2(b-a)}{n}, \ldots, a + \frac{(n-1)(b-a)}{n}, b\right\}$ be the partition of $[a, b]$ that equally divides the interval into n equal subintervals. Your problem is to show that $U(P_n, f) - L(P_n, f) = M\left(\frac{b-a}{n}\right)$. Try the case of $f(x) = 2x^2$ on $[0, 1]$. Use $n = 4$ and find a rectangle somewhere in the diagram whose area is the desired quantity and relate it to the difference of the upper and lower sum. Then generalize.

Exercise 9.10. Let $f(x) = \begin{cases} 0 & \text{if } x \text{ is rational} \\ 1 & \text{if } x \text{ is irrational} \end{cases}$. Let P be any partition of $[0,1]$. Compute $L(f,P)$ and $U(f,P)$. Compute $L(f)$ and $U(f)$. Is f integrable on $[0,1]$?

Exercise 9.11. Define $f(x)$ to be 1 for all x except $x = 0$ where $f(0) = 0$. Define $F(x) = \int_0^x f$ Clearly f is not continuous at $x = 0$. Show that F is differentiable at $x = 0$. Why doesnt this violate the second part of the Fundamental Theorem of Calculus (Theorem 9.27)?

Exercise 9.12. Let $a > 0$ be a real number and define f on $[-a, 3a]$ by $f(x) = |x|$. Explicitly find c (in terms of a) such that $f(c) = \frac{1}{4a}\int_{-a}^{3a} f(t)\ dt$.

Exercise 9.13. Let $f(x) = \begin{cases} 1 & \text{if } 0 \le x < 1 \\ 2 & \text{if } x = 1 \end{cases}$ and let $P = \{0, x_1, x_2, \ldots, x_{n-1}, 1\}$ be a partition of $[0,1]$. Compute $L(f,P)$ and $U(f,P)$. Your answers may contain terms from the partition.

Exercise 9.14. Let $f(x) = \begin{cases} 2x+1 & \text{if } 0 \le x \le 2 \\ 1-3x^2 & \text{if } 2 < x \le 4 \end{cases}$ and let $G(x) = \int_0^x f(t)\ dt$ for all x in $[0,4]$. Find $G(x)$ and determine where it is differentiable.

Exercise 9.15. Suppose that f is a continuous function on $[0,1]$ such that $f(x) \ge 0$ for all x in $[0,1]$. We know that f is integrable on the interval. Show that if $f(x) \ne 0$ at some x in $[0,1]$ then $\int_0^1 f > 0$.

Exercise 9.16. Let f and g be bounded functions on $[a,b]$. Further let

$$m_f = \text{glb}\{f(x) \mid x \in [a,b]\}, \qquad M_f = \text{lub}\{f(x) \mid x \in [a,b]\},$$
$$m_g = \text{glb}\{g(x) \mid x \in [a,b]\}, \text{ and} \qquad M_g = \text{lub}\{g(x) \mid x \in [a,b]\}.$$

Prove that $m_f + m_g \le \text{glb}\{f(x)+g(x) \mid x \in [a,b]\}$ and $\text{lub}\{f(x)+g(x) \mid x \in [a,b]\} \le M_f + M_g$.

Exercise 9.17. Let f and g be bounded functions on $[a,b]$. Let P be any partition of $[a,b]$. Prove that $L(f,P)+L(g,P) \le L(f+g,P)$ and $U(f+g,P) \le U(f,P)+U(g,P)$. (Hint: Use Exercise 8.17.)

Exercise 9.18. Find an example of a function f which is Riemann integrable on $[0,1]$ but does not satisfy the Mean Value Theorem for Integrals (Theorem 9.28). Explicitly show that it does not satisfy the theorem.

Exercise 9.19. Let $f(x) = x^3 + 2x$ on $[1,4]$. Let $\varepsilon = 0.05$. Find a partition P_ε such that $U(f, P_\varepsilon) - L(f, P_\varepsilon) < \varepsilon$.

Exercise 9.20. Let $f(x) = x^3$ on $[1,4]$. Let $P = \{1,2,3,4\}$ be a partition of $[1,4]$ and let C be the marking defined by applying the Mean Value Theorem (Theorem 8.16) to f over each of the subintervals of the partition. Show that $F(P,C,f') = \int_1^4 f'(x)\ dx$.

Chapter 10

What comes next in Real Analysis?

In this book we have started with an intuitive notion of the Real Numbers, a set of numbers used in Calculus, and have made that notion more precise. We have introduced convergence, continuity, differentiability, and integrability, concepts fundamental to Calculus and put them on a firmer foundation. Certainly Real Analysis does not end with chapter 9 of this book but is a thriving part of modern mathematics moving forward in many different directions. Here are some of those directions. This is not even close to an exhaustive view, but it does contains many of the areas that a second course in Real Analysis might study.

Defining the Real Numbers

We started with an intuitive view of the Real Numbers and declared that the Real Numbers is, as a set, a complete ordered field containing the Rational Numbers. The Completeness Axiom (Axiom 6) is an assumption about the Reals that allows us to fill up all the spaces between ratioanl numbers to create the reals. Again, it is an assumption, an axiom. In the mid 1800s two mathematiciains, Richard Dedekind (1831 1916) and Georg Cantor (1845 1918), defined the Real Numbers assuming the Rational Numbers exist and have their familiar axiomatic structure. Their definitions were quite different but resulted in the same set of Reals.

Dedekind defined the Reals via what he called cuts, pairs of sets of rational numbers. A cut is a partition of the rationals into two non-empty sets, A and B, such that their intersection is empty and their union is the set of rationals. Further every element of A is less than every element of B. Finally A does not have a largest element. Then Dedekind defines the four arithmetic operations on the set of all cuts and shows that this set is an ordered field. The Completeness Axiom then comes free of charge and Dedekind has defined the Real Numbers. This construction of the Reals from the Rationals is neither quick nor easy and it does not eliminate all mysteries from the Reals.

Cantor defined the Reals in terms of Cauchy sequences of rational numbers. One can define Cauchy sequences on the rationals simply by asking that the epsilon in the definition be a positive rational number. He then defines an equivalence relation on the set of all such Cauchy sequences. One way to look at this is to say that two Cauchy sequences are equivalent if the sequence that is created by interleaving the two given sequences is also Cauchy. He then, like Dedekind, proves that the set of all equivalence classes is what we think of as the Real numbers.

Uniform Convergence

In his textbook on Real Analysis, Augustin-Louis Cauchy (1789 1857) claimed to prove that the limit of a sequence of continuous functions is a continuous function. However, there is a very simple counter-example. For each natural number n, let $f_n : [0,1] \to [0,1]$ be defined by $f_n(x) = x^n$. Clearly each of these functions is continuous on $[0,1]$. However, if we fix an x and take the limit as n tends to infinity of $\{f_n(x)\}$ and call that limit $f(x)$, we have constructed a function discontinuous on $[0,1]$. Namely $f(x) = \lim_{n\to\infty} f_n(x) = \lim_{n\to\infty} x^n = \begin{cases} 0 & 0 \le x > 1 \\ 1 & x = 1 \end{cases}$. This follows from Theorem 5.14.(2). This is a somewhat disturbing situation. Once this phenomenon was discovered many other examples of it came to light and they showed that intuitive handling of infinite series and power series might lead to errors. Karl Weierstrass (1815 1897) defined the notion of Uniform Convergence of a sequence of functions and opened up the study of sequences and series of functions to deeper research. The kinds of problems that uniform convergence addressed were problems of the preservation of properties under convergence. For example if $\{f_n(x)\}$ is a sequence of functions on a common domain of A and the sequence has a point-wise limit of $f(x)$ on A, then does the limit process maintain continuity, differentiability, or Integrability?

Fourier Series

The concept of uniform convergence was very helpful in the study of what are called Fourier Series. Joseph Fourier (1768 1830) was a French engineer, and mathematician. He solved a version of what is known as the Heat Equation, a partial differential equation whose solution displays a steady-state temperature distribution on a surface. He solved the equation in a manner unique at the time, namely by using infinite series of trigonometric functions. He essentially replaced $a_n(x-c)^n$ in a traditional power series with terms like $a_n \sin(nx) + b_n \cos(nx)$. Using these series he could solve the partial differential equation at hand but his methods were met with deep skepticism (on convergence grounds). Today Fourier Series and later highly abstract generalizations of them (wavelets, etc.) are used to solve problems in the study and application of light and sound.

Lebesgue Integration

Bernhard Riemann (1826 1866) successfully gave an abstract definition of Integrability, indeed the definition presented in this text. Due to his work it is known as Riemann Integrability. However, as Riemann saw, there are functions which are not Riemann Integrable. We have looked at the example of the Dirichlet function on $[0,1]$ which is 1 on the irrationals and 0 on the rationals. The problem with this function is that it has so many discontinuities. With the developing study of infinite sets, a new approach to integration was imagined by Henri Lebesgue (1875 1941) and others. In this form of integration the Dirichlet function above is Integrable (Lebesgue Integrable) and has integral equal to 1. The Lebesgue integral agrees with the Riemann Integral for all functions where both exist but there are many functions which are Lebesgue Integrable but not Riemann Integrable. The real power of the Lebesgue version is that it avoids many of the problems uniform convergence was intended to address and gives a clearer path to the integrals of a limit of a sequence of Lebesgue Integrable functions.

One way to to see the difference between Riemann Integration and Lebesgue Integration is to note that the sets over which the Riemann Integral is defined are closed intervals, $[a, b]$, while for the Lebesgue Integral the basic sets are called measurable (or Lebesgue measurable) sets. In $[0,1]$

the subsets of the rationals and the irrationals are both measureable sets, the rationals having measure 0 while the irrationals have measure 1. Measure is essentially the length or more precisely the one-dimensional volume of a set.

Chapter 11

Supplemental Exercises

Exercise 11.1. The sequence of Fibonacci Numbers, $\{F_n \mid n = 1, 2, 3, \dots\}$ is defined by $F_1 = F_2 = 1$, $F_{n+2} = F_{n+1} + F_n$ for all $n \geq 1$. Prove that F_{3n} is an even number for every natural number n.

Exercise 11.2. Find the rational number, $\frac{p}{q}$, with $q \leq 10$, that is closest to $\sqrt{2}$.

Exercise 11.3. Suppose that a set A has 5 elements and a set B has 6 elements. Further suppose that their union has 8 elements. How many elements are there in the set $A \times B$?

Exercise 11.4. Let $f : \mathbb{R} \to \mathbb{R}$ be defined by $f(x) = 1 - x\sqrt{2}$. Find each of the following sets.

a) $f((-1, 1))$

b) $f((\sqrt{2}, \sqrt{3}))$

c) $f^{-1}((0, 1))$

d) $f((-1, \sqrt{5}))$

Exercise 11.5. Let $f : A \to B$ be a function. Prove or disprove:

a) If $U \subset A$ and $V \subset A$ then $f(U \cap V) = f(U) \cap f(V)$.

b) If $U \subset B$ and $V \subset B$ then $f^{-1}(U \cup V) = f^{-1}(U) \cup f^{-1}(V)$.

Exercise 11.6. Let $E = \{2, 4, 6, 8, 10, \dots\}$ and let $3\mathbb{Z} = \{\dots, -9, -6, -3, 0, 3, 6, 9, \dots\}$. Find an explicit bijection, f, from $3\mathbb{Z}$ to E, that is a one-to-one and onto function $f : 3\mathbb{Z} \to E$. Finding an explicit bijection means actually finding a formula for such an f and then proving that it is indeed a bijection.

Exercise 11.7. Let $A = (-3, 1)$. Prove that $-3 = \text{glb}(A)$.

Exercise 11.8. Let X be a set of real numbers and assume that u is both an element of X and an upper bound of X. Prove that $u = \text{lub}X$.

Exercise 11.9. Let a, b be distinct irrational numbers with $a < b$. Prove that there is an irrational number c satisfying $a < c < b$.

Exercise 11.10. In the Nested Intervals Theorem (Theorem 3.11) all intervals are closed intervals. Find an example of a nested sequence of non-empty open intervals such that their intersection is the empty set.

Exercise 11.11. Let the sequence $\{a_n\}$ be defined by $a_1 = 0$, $a_{n+1} = \frac{a_n}{3} - 2$ for all $n \in \mathbb{N}$. Prove that this sequence is bounded below. (Hint: First find a possible lower bound and then use induction.)

Exercise 11.12. Prove that $\lim_{n\to\infty} \frac{1}{n^2+1} = 0$.

Exercise 11.13. Suppose that r is a real number and $\lim_{n\to\infty} a_n = L$. Prove that $\lim_{n\to\infty} ra_n = rL$.

Exercise 11.14. Prove that $\lim_{n\to\infty} \left(1 + \frac{(-1)^n}{n}\right) = 1$.

Exercise 11.15. Suppose that $\lim_{n\to\infty} a_n = L$ and $\lim_{n\to\infty} b_n = M$. Prove that $\lim_{n\to\infty}(a_n - b_n) = L - M$.

Exercise 11.16. Let $a_1 = 1$, $a_{n+1} = \left(\frac{2}{3}\right) a_n + 7$. Prove that $\{a_n\}$ converges and find the limit of the sequence.

Exercise 11.17. Find an example of each of the following (if possible).

a) A bounded sequence that is divergent.

b) A convergent sequence that is not bounded.

c) An increasing sequence that is divergent.

d) A convergent sequence that is neither increasing nor decreasing.

e) An increasing sequence that is bounded below and divergent.

Exercise 11.18. For each of the following, find an example of a pair of sequences satisfying the stated conditions (if possible).

a) Sequences $\{a_n\}$ and $\{b_n\}$ are both divergent but $\{a_n + b_n\}$ converges.

b) Sequences $\{a_n\}$ and $\{b_n\}$ are both divergent but $\{a_n b_n\}$ converges.

c) Sequences such that $\{a_n\}$ converges, $\{b_n\}$ diverges and $\{a_n b_n\}$ converges.

d) Sequences such that $\{a_n\}$ converges, $\{b_n\}$ diverges and $\{a_n b_n\}$ diverges.

Exercise 11.19. Give an example (if one exists) of:

a) a Cauchy sequence of rational numbers that converges to an irrational number, and

b) a Cauchy sequence of rational numbers that does not converge.

Exercise 11.20. Show that for all x in the set $\{x \in \mathbb{R} \mid |x| < a\}$, the sum $\sum_{k=0}^{\infty} x^k > 0$.

Exercise 11.21. Let r be a real number and let $\sum_{k=1}^{\infty} a_k = A$. Prove that $\sum_{k=1}^{\infty} ra_k = rA$.

Exercise 11.22. Either find two real numbers, x and y, such that $\sum_{k=0}^{\infty} x^k$ and $\sum_{k=0}^{\infty} y^k$ both converge and satisfy $\left(\sum_{k=0}^{\infty} x^k\right)\left(\sum_{k=0}^{\infty} y^k\right) = \sum_{k=0}^{\infty} x^k y^k$ or explain why this is not possible.

Exercise 11.23. Which of the following series converge? If the series converges, find its sum.

a) $\sum_{k=0}^{\infty} \left(\frac{-3}{8}\right)^k$

b) $\sum_{k=0}^{\infty} \left(\frac{-8}{3}\right)^k$

c) $\sum_{k=0}^{\infty} \left(\frac{1}{3}\right)^k$

d) $\sum_{k=0}^{\infty} \frac{2^k}{3^{k+1}}$

e) $\sum_{k=0}^{\infty} \frac{1}{n^k}$ where n is a natural number greater than 1

Exercise 11.24. Let $\sum_{k=1}^{\infty} a_k$ and $\sum_{k=1}^{\infty} b_k$ be infinite series with $\sum_{k=1}^{\infty} a_k$ convergent. Suppose that $\{s_n\}$ is the sequence of nth partial sums of $\{a_n\}$ and $\{t_n\}$ is the sequence of nth partial sums of $\{b_n\}$. Finally assume that for each n, $t_n = s_n + \frac{1}{n}$. Determine whether $\sum_{k=1}^{\infty} b_k$ converges or diverges and justify your answer.

Exercise 11.25. Let $s_n = \frac{n+1}{2n-1}$ be the nth partial sum of the sequence $\{a_n\}$. Find a_{100} and $\sum_{k=1}^{\infty} a_k$.

Exercise 11.26. a) Prove that $(-\infty, -2)$ is an open set.

b) Prove that $[-3, 4]$ is a closed set.

Exercise 11.27. Prove that if A is a set that has a limit point then A is an infinite set.

Exercise 11.28. Prove that -2 is a limit point of $(-4, -2)$ using the theorem on sequences and limit points (Theorem 6.17).

Exercise 11.29. Prove from the definition that $\{1\}$ is a compact set.

Exercise 11.30. Prove that $\lim_{x \to 2} 1 - 2x = -3$.

Exercise 11.31. Prove that $\lim_{x \to -2} 2x^2 - 1 = 7$.

Exercise 11.32. Let $f(x) = \frac{1}{x}$ with domain $(-\infty, 0) \cup (0, \infty)$. Show that $\lim_{x \to 0} f(x)$ does not exist. (Hint: Consider the sequence $\{x_n\} = \left\{\frac{1}{n}\right\}$ and assume that $\lim_{x \to 0} f(x) = L$ for some real number L.)

Exercise 11.33. Let $f : A \to \mathbb{R}$, c be a limit point of A, and $\lim_{x \to c} f(x) = 1$. Prove that there is an $\varepsilon > 0$ such that $f(x) > 0$ for all $x \neq c$, $x \in N_\varepsilon(c) \cap A$.

Exercise 11.34. Let K be a non-empty compact set of real numbers. Prove that lub$K$ and glb$K$ both exist and are in K.

Exercise 11.35. Let $A = \left\{1, \frac{1}{2}, \frac{1}{3}, \frac{1}{4}, \ldots, 0\right\}$ and let $f : A \to \mathbb{R}$ be defined by $f(x) = \frac{1}{x}$. At what points is f continuous?

Exercise 11.36. Let $f : (-4, -1) \to \mathbb{R}$ be defined by $f(x) = \frac{1}{(x+1)^2}$. Show that f is not uniformly continuous on $(-4, -1)$.

Exercise 11.37. Let $f : [0, \infty) \to \mathbb{R}$ be defined by $f(x) = \sqrt{x}$. Show that f is not uniformly continuous on $(-4, -1)$.

Exercise 11.38. Using the limit definition of the derivative at a point $x = c$, namely $f'(c) = \lim_{x\to c} \frac{f(x)-f(c)}{x-c}$, find the derivative of $f(x) = \sqrt{2x}$ at $c = 8$.

Exercise 11.39. Given a function $f(x)$, define $F(x) = \lim_{h\to 0} \frac{f(x+h)-f(x-h)}{2h}$.

a) Show that if $f(x) = x^2$ then $F(x) = f'(x)$.

b) Compute $F(0)$ if $f(x) = |x|$. Does $F(x) = f'(x)$ in this case?

Exercise 11.40. Let $f(x) = x^2$ and $c = 1$. Note that $f'(c) = 2 > 0$. Find a $\delta > 0$ such that $f(c-h) < f(c) < f(c+h)$ for all $0 < h < \delta$.

Exercise 11.41. Let $f : \mathbb{R} \to \mathbb{R}$ be defined by $f(x) = \begin{cases} 1 & \text{if } x \text{ is rational} \\ 0 & \text{if } x \text{ is irrational} \end{cases}$. Let c be any real number. Show that $\lim_{x\to c} f(x)$ does not exist.

Exercise 11.42. Let $f : \mathbb{R} \to \mathbb{R}$ be defined by $f(x) = \begin{cases} x & \text{if } x \text{ is rational} \\ 0 & \text{if } x \text{ is irrational} \end{cases}$. Show that $\lim_{x\to 0} f(x)$ exists and equals $f(0)$, hence that f is continuous at $x = 0$. Show that $\lim_{x\to c} f(x)$ does not exist for any other value of x.

Exercise 11.43. Verify the Mean Value Theorem for $f(x) = \sqrt{x}$ over the interval $[1, 9]$.

Exercise 11.44. Let $f : \mathbb{R} \to \mathbb{R}$ be defined by $f(x) = \begin{cases} x^3 & x < 0 \\ -x^2 & 0 \le x \end{cases}$.

a) Is f continuous at $x = 0$? Why?

b) Is f differentiable at $x = 0$? What is $f'(0)$?

c) Verify the Mean Value Theorem for $f(x)$ over $[-2, 1]$.

Exercise 11.45. Prove that $\lim_{x\to 0} \cos\left(\frac{1}{x}\right)$ does not exist.

Exercise 11.46. Let $f(x) = x^2 + 3x$ on $[0, 4]$ and let d be between 0 and 28, that is $0 = f(0) < d < 28 = f(4)$. Find c in $(0, 4)$ such that $f(c) = d$. Note this does not mean pick a particular value for d, but solve the problem for an arbitrary d.

Exercise 11.47. Prove that $\{1, 2\}$ is disconnected.

Exercise 11.48. Let f be defined on $[0, 2]$ by the following formula:

$$f(x) = \begin{cases} x & \text{if } x \text{ is irrational} \\ \sqrt{3} & \text{if } x \text{ is rational} \end{cases}.$$

Let P be the partition $\{0, 1, \sqrt{2}, 2\}$. Compute $L(P, f)$ and $U(P, f)$.

Exercise 11.49. Find a partition, P_ε, on $[0, 2]$ such that $U(P_\varepsilon, x^2 + x) - L(P_\varepsilon, x^2 + x) < \varepsilon = 0.02$.

Exercise 11.50. Let $f : [0, 1] \to \mathbb{R}$ be defined by $f(x) = \begin{cases} 1 & \text{if } 0 \le x < 1 \\ 2 & \text{if } x = 1 \end{cases}$ and let $P = \{0, x_1, x_2, \ldots, x_{n-1},$ be a given partition of $[0, 1]$. Compute $U(P, f) - L(P, f)$.

Bibliography

[1] Sarah-Marie Belcastro and Michael Green. The Cantor Set Contains $\frac{1}{4}$? Really? *The College Mathematics Journal*, **32**(1):55–56, January 2001.

[2] Georg Cantor. Über eine elementare Frage de Mannigfaltigketslehre. *Jahresber. Deutsch. Math.-Verein.*, **1**:75–78, 1890-1891.

[3] John Franks. Cantor's Other Proofs that $\mathbb{R}$ is Uncountable. *Mathematics Magazine*, **83**(4):283–289, October 2010.

[4] Casper Goffman. Completeness of the Real Numbers. *Mathematics magazine*, **47**(1):1 – 8, January 1974.

[5] Thomas L. Heath and Euclid. *The Thirteen Books of Euclids Elements, Books 1 and 2.* Dover Publications, Inc., USA, 1956.

[6] Ioana Mihaila. Cantor, $\frac{1}{4}$, and Its Family and Friends. *The College Mathematics Journal*, **33**(1):21–23, January 2002.

[7] James Propp. Real analysis in reverse. *The American Mathematical Monthly*, 120, 04 2012.

[8] J. Stewart. *Calculus: Early Transcendentals.* Cengage Learning, 2015.

[9] Holger Teismann. Toward a more complete list of completeness axioms. *American Mathematical Monthly*, 120, 02 2013.

List of Definitions and Axioms

Alphabetical Index

Made in the USA
Monee, IL
22 July 2026

57854874R00066